# Innovation, Technology, and Knowledge Management

**Series editor**
Elias G. Carayannis
School of Business
George Washington University
Washington, D.C., USA

More information about this series at http://www.springer.com/series/8124

Martin Curley • Bror Salmelin

# Open Innovation 2.0

The New Mode of Digital Innovation
for Prosperity and Sustainability

 Springer

Martin Curley
Innovation Value Institute
Maynooth University
Maynooth, Kildare, Ireland

Bror Salmelin
DG Communications Networks
Contents and Technology, EU Commission
Brussels, Belgium

ISSN 2197-5698          ISSN 2197-5701   (electronic)
Innovation, Technology, and Knowledge Management
ISBN 978-3-319-62877-6        ISBN 978-3-319-62878-3   (eBook)
DOI 10.1007/978-3-319-62878-3

Library of Congress Control Number: 2017948655

Printed on acid-free paper

This Springer imprint is published by Springer Nature
The registered company is Springer International Publishing AG
The registered company address is: Gewerbestrasse 11, 6330 Cham, Switzerland

# Foreword

Two key words during my Presidency of the European Committee of the Regions (CoR) have been innovation and digitalization. The world around us and our conception of it are changing with accelerating speed. The new major success factor not only for industry but also for cities and regions is speed or better said velocity. Dramatic changes are posing both practical and exceptional theoretical and systemic challenges. The paradigms are changing from industrial society through information society to knowledge and innovation society so profoundly and quickly that it is hard to keep pace with them and make sense of the unparalleled transformations.

This changing landscape forms an excellent frame and reasoning to read this book, written by Professor Martin Curley and Bror Salmelin. Europe needs renewal through a new entrepreneurial mind-set and digitalization.

I am convinced that a Digital Europe based on in-depth bench learning and partnerships between cities and regions is becoming a reality. The CoR has challenged all the cities and regions in Europe to take stronger actions in becoming forerunners, especially in tackling societal challenges and in creating sustainable growth and new jobs. Let us learn what recent industrial and public sector practice has to offer for the European renewal.

Let us speed up the digital transformation by integrating the industrial experiences with the evidence-based knowledge, i.e., best practices and concepts, to operate via European digitalized open innovation platforms and thus getting new European innovations faster to the global markets.

The learnings can be extended beyond Europe, and in a new VUCA (volatility, uncertainty, complexity, and ambiguity) world, extraordinary leadership is called for to help guide us all to a better place. Digital technologies form the essential foundation for inventing the future. This book sheds light on the path to how we can collectively both simultaneously drive economic growth and improve society in a sustainable way.

Creating an understanding of the nature of disruptive change is the driver for providing sustainable benefits for society and global businesses. Let me encourage you to read and learn from what Martin Curley and Bror Salmelin have written. This book can be a strong push forward in your personal knowledge sharing and co-creation process.

European Committee of the Regions                        Markku Markkula
Bruxelles, Belgium

# Dedication and Acknowledgments

This book is a synthesis of much work, research, and experience from various innovation landscapes, and we thank all who have contributed.

Martin:

To my family for all their support for which I am very grateful,

For surgeons Brian Mehigan and Donal Maguire and their colleagues for their brilliant work and care,

To the memory of Pauline Carbury and Alice Flanagan, a lovely woman and a lovely child who both left this world too soon,

For the OI2 community for their energy and creativity in helping make a difference.

Bror:

This journey in innovation over the years has been supported by my family whom I thank wholeheartedly.

Inspiration has also been given from numerous discussion partners and friends reflecting the thoughts and encouraging to go further.

*"Live life out of your imagination, not your history"*

Stephen Covey

# Series Foreword

The Springer book series *Innovation, Technology, and Knowledge Management* was launched in March 2008 as a forum and intellectual, scholarly "podium" for global/local, transdisciplinary, trans-sectoral, public–private, and leading/"bleeding"-edge ideas, theories, and perspectives on these topics.

The book series is accompanied by the Springer *Journal of the Knowledge Economy*, which was launched in 2009 with the same editorial leadership.

The series showcases provocative views that diverge from the current "conventional wisdom," that are properly grounded in theory and practice, and that consider the concepts of *robust competitiveness*,[1] *sustainable entrepreneurship*,[2] and *democratic capitalism*,[3] central to its philosophy and objectives. More specifically, the aim of this series is to highlight emerging research and practice at the dynamic intersection of these fields, where individuals, organizations, industries, regions, and nations are harnessing creativity and invention to achieve and sustain growth.

---

[1] We define *sustainable entrepreneurship* as the creation of viable, profitable, and scalable firms. Such firms engender the formation of self-replicating and mutually enhancing innovation networks and knowledge clusters (innovation ecosystems), leading toward robust competitiveness (E.G. Carayannis, *International Journal of Innovation and Regional Development* 1(3). 235–254, 2009).

[2] We understand *robust competitiveness* to be a state of economic being and becoming that avails systematic and defensible "unfair advantages" to the entities that are part of the economy. Such competitiveness is built on mutually complementary and reinforcing low-, medium-, and high-technology and public and private sector entities (government agencies, private firms, universities, and nongovernmental organizations) (E.G. Carayannis. *International Journal of Innovation and Regional Development* 1(3). 235–254. 2009).

[3] The concepts of *robust competitiveness* and *sustainable entrepreneurship* are pillars of a regime that we call *"democratic capitalism"* (as opposed to "popular or casino capitalism"), in which real opportunities for education and economic prosperity are available to all, especially—but not only—younger people. These are the direct derivative of a collection of top-down policies as well as bottom-up initiatives (including strong research and development policies and funding, but going beyond these to include the development of innovation networks and knowledge clusters across regions and sectors) (E.G. Carayannis and A. Kaloudis. *Japan Economic Currents*. p. 6–10 January 2009).

Books that are part of the series explore the impact of innovation at the "macro" (economies, markets), "meso" (industries, firms), and "micro" levels (teams, individuals), drawing from related disciplines such as finance, organizational psychology, research and development, science policy, information systems, and strategy, with the underlying theme that for innovation to be useful it must involve the sharing and application of knowledge.

Some of the key anchoring concepts of the series are outlined in the figure below and the definitions that follow (all definitions are from E.G. Carayannis and D.F.J. Campbell, *International Journal of Technology Management,* 46, 3–4, 2009).

Conceptual profile of the series *Innovation, Technology, and Knowledge Management*

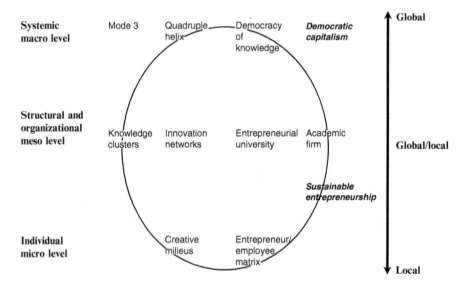

- The "Mode 3" Systems Approach for Knowledge Creation, Diffusion, and Use: "Mode 3" is a multilateral, multinodal, multimodal, and multilevel systems approach to the conceptualization, design, and management of real and virtual, "knowledge-stock" and "knowledge-flow," modalities that catalyze, accelerate, and support the creation, diffusion, sharing, absorption, and use of cospecialized knowledge assets. "Mode 3" is based on a system-theoretic perspective of socioeconomic, political, technological, and cultural trends and conditions that shape the coevolution of knowledge with the "knowledge-based and knowledge-driven, global/local economy and society."
- Quadruple Helix: Quadruple helix, in this context, means to add to the triple helix of government, university, and industry a "fourth helix" that we identify as the "media-based and culture-based public." This fourth helix associates with "media," "creative industries," "culture," "values," "lifestyles," "art," and perhaps also the notion of the "creative class."

- Innovation Networks: Innovation networks are real and virtual infrastructures and infratechnologies that serve to nurture creativity, trigger invention, and catalyze innovation in a public and/or private domain context (for instance, government–university–industry public–private research and technology development coopetitive partnerships).
- Knowledge Clusters: Knowledge clusters are agglomerations of cospecialized, mutually complementary, and reinforcing knowledge assets in the form of "knowledge stocks" and "knowledge flows" that exhibit self-organizing, learning-driven, dynamically adaptive competences and trends in the context of an open systems perspective.
- Twenty-First Century Innovation Ecosystem: A twenty-first century innovation ecosystem is a multilevel, multimodal, multinodal, and multiagent system of systems. The constituent systems consist of innovation metanetworks (networks of innovation networks and knowledge clusters) and knowledge metaclusters (clusters of innovation networks and knowledge clusters) as building blocks and organized in a self-referential or chaotic fractal knowledge and innovation architecture (Carayannis 2001), which in turn constitute agglomerations of human, social, intellectual, and financial capital stocks and flows as well as cultural and technological artifacts and modalities, continually coevolving, cospecializing, and cooperating. These innovation networks and knowledge clusters also form, reform, and dissolve within diverse institutional, political, technological, and socioeconomic domains, including government, university, industry, and non-governmental organizations and involving information and communication technologies, biotechnologies, advanced materials, nanotechnologies, and nextgeneration energy technologies.

*Who is this book series published for?* The book series addresses a diversity of audiences in different settings:

1. *Academic communities:* Academic communities worldwide represent a core group of readers. This follows from the theoretical/conceptual interest of the book series to influence academic discourses in the fields of knowledge, also carried by the claim of a certain saturation of academia with the current concepts and the postulate of a window of opportunity for new or at least additional concepts. Thus, it represents a key challenge for the series to exercise a certain impact on discourses in academia. In principle, all academic communities that are interested in knowledge (knowledge and innovation) could be tackled by the book series. The interdisciplinary (transdisciplinary) nature of the book series underscores that the scope of the book series is not limited a priori to a specific basket of disciplines. From a radical viewpoint, one could create the hypothesis that there is no discipline where knowledge is of no importance.
2. *Decision makers—private/academic entrepreneurs and public (governmental, subgovernmental) actors:* Two different groups of decision makers are being addressed simultaneously: (I) private entrepreneurs (firms, commercial firms, and academic firms) and academic entrepreneurs (universities), interested in optimizing knowledge management and in developing heterogeneously

composed knowledge-based research networks; and (2) public (governmental, subgovernmental) actors that are interested in optimizing and further developing their policies and policy strategies that target knowledge and innovation. One purpose of public *knowledge and innovation policy* is to enhance the performance and competitiveness of advanced economies.

3. *Decision makers in general:* Decision makers are systematically being supplied with crucial information, for how to optimize knowledge-referring and knowledge-enhancing decision-making. The nature of this "crucial information" is conceptual as well as empirical (case study-based). Empirical information highlights practical examples and points toward practical solutions (perhaps remedies); conceptual information offers the advantage of further-driving and further-carrying tools of understanding. Different groups of addressed decision makers could be decision makers in private firms and multinational corporations, responsible for the knowledge portfolio of companies; knowledge and knowledge management consultants; globalization experts, focusing on the internationalization of research and development, science and technology, and innovation; experts in university/business research networks; and political scientists, economists, and business professionals.

4. *Interested global readership:* Finally, the Springer book series addresses a whole global readership, composed of members who are generally interested in knowledge and innovation. The global readership could partially coincide with the communities as described above ("academic communities," "decision makers") but could also refer to other constituencies and groups.

Elias G. Carayannis

# Preface

*The way is long if one follows precepts (rules); the way is short if one follows patterns.*
  Seneca.

We are at a unique point in time where we have multiple disruptive technologies all showing up at the same time, creating a chain reaction of disruptive change. In this perfect storm, organizations and indeed ecosystems have a choice to react and let change happen or to proactively try to invent and innovate better outcomes. Open Innovation 2.0 (OI2) is the new paradigm and methodology for Digital Innovation. A new primordial soup exists which is bound by digital, enabled by digital, and fueled by digital where all actors in business and society have the opportunity to quickly create transformation solutions using agile methods. Based on our research and practice, we share the first version of an OI2 pattern language including core patterns to help innovators across the spectrum to increase the probability of success using a Digital platform and ecosystem approach. We have distilled these first patterns as we have observed the signals emerge from the noise in the rapidly exploding field of digital innovation. We present these initial patterns as a minimum viable platform (MVP) for OI2-led digital innovation, knowing instantly that almost before the ink is dry upon printing some of these will need to change as we learn and as dynamics change. We present the MVP OI2 pattern language to provide a rudimentary taxonomy and vocabulary to allow practitioners experiment and test these patterns with real-life projects and to give a base platform for researchers and practitioners to help expand and more fully describe the OI2 pattern language. Using the agile and rapid experimentation approach, we hope and expect that the OI2 pattern language will be iterated and improved quickly providing transformational value to governments, industry, academics, and citizens/users alike. Again using OI2 principles, we provide a "good enough" first version of the core patterns knowing already that there are omissions/errors rather than waiting for a much more polished version delivered later. We welcome your feedback and hope the book and associated body of knowledge are helpful to you.

Maynooth, Ireland                                                                    Martin Curley
Brussels, Belgium                                                                    Bror Salmelin
15 June 2017

# Contents

# About the Authors

**Martin Curley** is Professor of Innovation at Maynooth University, Ireland. He is co-founder of the Innovation Value Institute, an industry-academic open innovation consortium that strives to research and promote structural change in the way companies and governments achieve value through information technology. He chairs the European Union Open Innovation Strategy and Policy group (OISPG), an industry-led group advising on strategic priorities for open and service innovation and is a member of the EU Connect Advisory Forum and the EU Horizon 2020 Advisory Group on international Cooperation. Martin is also Senior Vice President and Group Head for Global Digital Practice at MasterCard providing digital thought and practice leadership to MasterCard customers. Previously he was vice president at Intel Corporation and director of Intel Labs Europe, the company's network of more than 40 research labs, development centers and open innovation collaborations spanning the European region. He also served as a senior principal engineer at Intel Labs Europe and lead Intel's research and innovation engagement with the European Commission and the broader European Union research ecosystem.

Before assuming his current position in 2009, Curley was global director of IT innovation at Intel. Earlier in his Intel career, he held a number of senior IT management and automation positions for Intel in the United States and Europe. Before joining Intel in 1992, he held management and research positions at General Electric in Ireland and at Philips Electronics in the Netherlands. Curley is the author or co-author of five books and dozens of papers on technology management for value, innovation and entrepreneurship. He is a Member of the Royal Irish Academy, fellow of the Institution of Engineers of Ireland, the British Computer Society, the Irish Computer Society and the Irish Academy of Engineering. Martin was previously a visiting scholar at MIT Sloan Centre for Information Systems Research and was awarded joint European Chief Technology Officer of the year for 2015/2016.

**Bror Salmelin** is Advisor for Innovation Systems, DG Communications Networks, Contents and Technology at the European Commission. He was the deputy of the IT department of the Finnish Technology and Innovation Agency in 1984-97. In this capacity he represented Finland in the EU ESPRIT and ICT research programmes. He was the EFTA chair and co-designer of the global Intelligent Manufacturing Systems initiative since 1990. He was representing Finland in the Consulate of Los Angeles as Vice Consul in 1997-98 with the responsibility to build bridges between the Finnish and Californian innovation systems and moved in 1998 to European Commission to lead the units of Integration in Manufacturing, later eCommerce and Collaborative work before becoming advisor to the DG. He is the initiator of the European Network of Living Labs which now has more than 350 sites worldwide, and also the initiator of the Open Innovation activities in the European Commission. He is member of the New Club of Paris, IVI advisory board and founder of the EU OISPG.

# Chapter 1
# Introduction

*'The dogmas of the quiet past are inadequate to the stormy present'.*

*Abraham Lincoln*

Many people recognize that innovation is not just an imperative for economic and social progress but that it is an art and skill, which underpins progress and survival of the human species. We have all seen how industries such as the music and book industry, which existed relatively unchanged for decades, have been transformed through digital technologies. These changes have exemplified Schumpeter's 'creative disruption' and Christensen's 'disruptive technologies' where new players such as Amazon and Spotify have replaced incumbents such as HMV and Borders stores. As a global society, we are ready for the next stage of disruptive change as societal level systems such as those for smart cities, agriculture, energy, health, and transportation systems are set for digital disruption. Equally, many industries are ripe for digital disruption. The potential benefits are enormous and so also are the challenges. The innovations and changes which will be required to drive these transformations will require much collaboration and alignment across ecosystems and indeed society. The emerging paradigm of Open Innovation 2.0 (OI2) (Curley and Salmelin 2013/2014; Curley and Formica 2013; Curley 2016; Madelin 2016) offers a series of design patterns to help innovators move efficiently and grasp this new opportunity of digital.

OI2 is a paradigm based on principles of integrated collaboration, co-created shared value, cultivated innovation ecosystems, unleashed exponential technologies, and rapid adoption due to network effects. We believe that innovation can be a discipline practised by many, rather than an art mastered by few. In addition, OI2 asserts that the probability of breakthrough improvements increases as a function of diverse multidisciplinary experimentation.

OI2 is both enabled by and fuelled by Digital, so that a virtuous cycle of innovation is enabled with each digital innovation providing value and becoming an infrastructure for future innovations to leverage.

© Springer International Publishing Switzerland 2018
M. Curley, B. Salmelin, *Open Innovation 2.0*, Innovation, Technology, and Knowledge Management, DOI 10.1007/978-3-319-62878-3_1

## 1.1   Simultaneous Arrival of Multiple Disruptive Technologies

We are potentially witnessing the biggest change in the history of the planet and its population. In the past, one disruptive technology such as the internal combustion engine or Tesla's system of alternating current generators, motors, and transformers enabled systematic electrification, which drove a wave of industry, economic, and societal change. Today, we have multiple disruptive technologies all arriving at the same time, which is resulting in the opportunity and inevitability of exponential change.

Cloud computing has changed the face of computing, making anytime, anywhere, on-demand computing available so that a small start-up in Malaysia or Sweden has access to the same kind of computing resources that previously only a General Electric or SAP could afford. This phenomenon is dramatically lowering the barriers and cost of entry to global opportunity, with an example 40× cost difference between procuring the service from a public cloud on a Microsoft Azure platform compared to provisioning one's own system. There are already examples of whereby the use of Digital enables a reduction of 95% of cost for a financial transaction conducted online compared to in a bank branch.

Similarly, big data and data-driven innovation is creating significant information monetization opportunities. Former European Research Commissioner, Maire Geogeghan Quinn, coined the phrase 'Knowledge is the crude oil of the 21st century' and it aptly describes the opportunity. For many individuals, organizations and ecosystems the emergence of big data, with ever more powerful machines and data mining and machine learning algorithms will present an opportunity of a lifetime. However, the opportunity of a lifetime has to be taken in the lifetime of the opportunity.

The Internet of Things (IOT) is poised to be the largest industry in the history of electronics and yet this impact will be dwarfed by the impact IOT will have in transforming many industries and societal systems. Autonomous driving is but one example of widespread adoption and impact of IOT.

## 1.2   Evolution or Revolution? A New Paradigm

'All truth passes through three stages, first it is ridiculed, and second it is violently opposed. Third, it is accepted as self-evident'. Schopenhauer

Several streams of evolutionary thinking are brought together to the Open Innovation 2.0 paradigm for Digital Innovation. However, the shift from the old paradigms to OI2 requires a radical transformation of culture, organizations, and the innovation environments.

As stated in the report by OISPG (Open Innovation Strategy and Policy Group) in the publication 'Societal Impact of Open Service Innovation' the emergence of the reverse innovation pyramid means that the fundamentals of the innovation process have changed. The user has moved from being an adopter of innovations to often being a key contributor to the innovation process. We will elaborate on the consequences a bit later.

Using the definition of Science Historian Thomas Kuhn, a paradigm refers to a set of practices that define a scientific discipline at a particular point in time. Another definition of paradigm refers to a paradigm as a 'pattern or model, an exemplar'. According to Kuhn, the sciences alternate between two states 'normal science' and 'revolution'. We believe that the discipline of innovation is passing through a strategic inflection point to the paradigm of Innovation 2.0.

Important background thinking for open innovation comes from the theories of dynamic capabilities and holonic/fractal enterprises, as the dynamic configurability of resources and organizations becomes ever more important. How to make organizations and ecosystems at the same time agile, robust, and effective by sharing and multiplying the competencies and capacities. Emergence is a crucial concept for dynamic ecosystems and is the appearance of patterns or systems at the macro-level, which emerge or evolve from the interactions amongst elements in an ecosystem.

Last decade, Henry Chesbrough (2003) eloquently conceptualized the idea of open innovation where ideas can pass to and from different organizations and travel on different exploitations vectors for value creation. Unlike hype cycles around solutions such as Second Life it appears that, the term open innovation is not subject to the typical hype cycle associated with new concepts or technologies.

In this decade, we are witnessing a new level of open innovation and an increasing sophistication and complexity associated with innovation. However, still the thinking and theory about innovation lagged the practise with much of the dialogue and publications concern organizations cross-licensing and collaborating. Indeed also, the funnel described by Chesbrough was based on the linear innovation model, while we observe in practice that innovation today is very much an iterative, non-linear model.

In this context, we observe that a new mode of technical and societal innovation is emerging, with blurred lines between universities, industry, governments, and featuring users and indeed communities as innovators. For example in Brixton, UK a broad set of stakeholders including Lambeth Council, Transport for London, companies, and most importantly schools and children fused participatory design, data, and play to co-design urban services that approach sustainability through community sensing, data visualization, behaviour change, and ambient technology.

OI2 is a new mode of innovation based on principles of integrated collaboration, co-created shared value, cultivated innovation ecosystems, unleashed exponential technologies, experimentation and focus on adoption and sustainability. OI2 is rooted in a vision of sustainable intelligent living where smart solutions are developed and diffused meeting needs while being resource and environmentally efficient. OI2 also promises significant improvements in the pace, productivity, predictability, and profitability of our collective innovation efforts.

Figure 1.1 conceptualizes OI2 as a new primordial soup, which is bound, enabled, fuelled, and connected by advanced computing and communications infrastructure and other digital technologies. In this new milieu, everyone can choose to be an innovator as exemplified by the hundreds of thousands of app developers who now

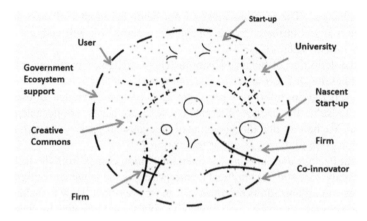

**Fig. 1.1**  Open Innovation 2.0—a new milieu

exist and contribute to various different ecosystems. In this new environment, the unit of competition has changed from the organization to the ecosystem and from the product to the platform.

## 1.3   Enabling Forces: A Perfect Storm

The collision of three mega trends Moore's law, mass collaboration, and sustainability, each of which are mutually reinforcing is creating a unique opportunity for us to leverage our collective intelligence and energies. Here, the nature of innovation from changes from a linear to non-linear process to drive innovation and deliver structural outcomes far beyond the scope of what anyone organization or individual could achieve on their own.

### 1.3.1   Moore's Law/Digital

We can define 'Digital' as innovation with and the use of *information* and *technology* to improve human, organization, and ecosystem performance and sustainability. Digital is the synthesis and synergy of information, silicon, network and software capabilities, and economics. Moore's law has become a proxy for the exponential advancement of capabilities in various information and communication-related domains. Moore's law was an observation that transistor density on integrated circuits would double every 2 years or so and that this would be delivered at less or equal cost. Moore's law became a competitive challenge, indeed an innovation strategy for the entire semiconductor industry to work together to ensure that the prediction was met. The resultant impact was that Moore's law became a driver of technological and social change, dramatically improving productivity and driving economic growth.

Significant technological innovation has ensured that Moore's law continues to hold true to essentially deliver the doubling of compute performance delivered at less or equal cost every two years or so. The innovation revolution is enabled by increasing levels of connectivity and catalysed by the emergence of exponential technologies such as Internet of Things, clouds, and open data. Thus, ordinary things such as dishwashers to cars become smart, connected, and collaborative. When smart things and indeed people are connected the intrinsic intelligence and our collective execution capability is multiplied exponentially.

Not only is there great opportunity to create and extract value particularly when data is shared, aggregated, and analysed across domains, a transformation opportunity exists to create new high frequency, high precision management control circuits in societal level systems, where previously only open loop control was possible. A simple example is a gully signalling to a city management system that it is blocked whereas a more complex example is a dynamic congestion-based charging system which automatically adjusts, changes traffic flow, and offers park and ride incentives based on parameters such levels of traffic and air quality in a city.

### 1.3.2    Mass Collaboration

The European Internet Foundation (EIF) have proposed for the next decades a paradigm of a world driven by mass collaboration, enabled by the ubiquitous availability of high speed, high capacity digital networks and services. EIF predicts the inexorable spread of *purpose-driven* online collaboration as the role of networks evolves from not just enabling communication and transactions but value creation through collaboration. We have all witnessed the phenomenon of *social production,* whereby people contribute to generate economic value, where there are little of no monetary incentives involved with the ongoing evolution of Wikipedia and the development of Linux being primary examples.

We will see mass people-to-people, machine-to-machine, and machine-to-people collaboration. Sometimes, this collaboration will be proactive creative collaboration where individuals as part of a community or as a part of a more formal innovation configurations will co-design and co-create solutions such as a new city services or transformation of an electrical grid. Other times, we will give permission to our devices to collaborate together to figure out an optimum solution to a given scenario, for example, real-time car-to-car communication and collaboration to determine the best sequencing of traffic at an upcoming junction to minimize transit times. The EU FP7 project TEAM (Total Elastic Adaptive Mobility) is focused on developing cooperative systems for energy efficient and sustainable mobility, with drivers, travellers, and infrastructure acting as a TEAM—adapting to each other and to the situation, creating optimized mobility conditions. The TEAM solutions were piloted in several European cities including Athens and Turin with players such as BMW, FIAT, Volvo, and Intel involved as well as naturally the municipalities and citizens.

## *1.3.3  Sustainability*

With the adoption of the new UN Sustainable Development goals, the recent Paris COP 21 agreement and the increasing trend of extreme weather events, individuals, and communities are becoming more sustainability focused. In parallel, there is a slow but growing recognition of the need to move from the 'take, make, dispose' mode of today's linear economy to a circular economy that preserves and enhances natural capital. The nirvana of sustainability is the ability to decouple growth from resource consumption and environmental impact and knowledge-driven entrepreneurship provides a potential pathway to achieve this. Former EU Research Commissioner Maire Geogeghan Quinn's statement that 'knowledge is the crude oil of the 21st century' aptly describes the opportunity. By leveraging the astonishing possibilities enabled by Moore's law, harnessing the collective intelligence and energy of people and machines worldwide through mass collaboration, focused on new solutions which are intrinsically sustainable we may be about to witness something akin to a pre-Cambrian explosion of impactful innovations.

Similar to the Gutenberg's invention of the printing press, the invention and evolution of modern computing and communications technology is a fundamental disruptor to the fabric and nature of society. We have all witnessed how industries such as the music and book selling industries have been transformed through ICT led by companies such as Apple and Amazon, respectively. This is Schumpeter's 'creative destruction' at work or this could alternatively be termed 'Digital Darwinism'. However, the next phase of Digital transformations will deliver significantly more value, will be orders of magnitude more difficult, and will require significant citizen involvement to maximize the chances of success. Transformation of our cities, energy grids, and healthcare systems will ultimately evolve through a process of emergence; however, the opportunity exists to proactively take charge and move much more quickly to the benefits promised by these transformations. OI2 is an emerging innovation mode, which is based on an evolving set of design patterns, i.e. general reusable solutions to commonly occurring problems, which can accelerate the delivery of innovation benefits.

## 1.4  Innovation Modes

These three driving forces have enabled a new mode of innovation and the following table describes key characteristics of the new mode compared to previous modes of innovation (Fig. 1.2).

The innovation landscape is changed; we have moved from the linear innovation model to something much more complex; a parallel and even mash-up type of innovation models where collisions of disciplines and ideas ignite innovation.

**HOW INNOVATION MODES HAVE EVOLVED**

| Closed innovation | Open innovation | Open innovation 2.0 |
|---|---|---|
| Dependency | Independency | Interdependency |
| Subcontracting | Cross-licensing | Cross-fertilization |
| Solo | Bilateral | Ecosystem |
| Linear | Linear, leaking | Nonlinear mash-up |
| Linear subcontracts | Bilateral | Triple or quadruple helix |
| Planning | Validation, pilots | Experimentation |
| Control | Management | Orchestration |
| Win–lose game | Win–win game | Win more–win more |
| Box thinking | Out of the box | No boxes! |
| Single entity | Single discipline | Interdisciplinary |
| Value chain | Value network | Value constellation |

**Fig. 1.2**  Innovation modes

## 1.4.1   *Dependency > Independency > Interdependency*

In linear enterprise behaviour, the dependencies are very dominant when we, e.g. think about supply chains where designs and processes form a controlled, well-defined sequence of operations. If some stage goes wrong, the whole process is halted until the very same problem is solved. This leads also to long planning times for modifications as the system has no redundancy. Independency often means that the company or cluster has increased internal redundancy for its operations. In Open Innovation 2.0, the ecosystem builds strong interdependencies while also, the entities become autonomous and processes diversified towards many supplies and vendors. This is important also in all phases of the innovation where ideas flow freely, ignite new ones and are elaborated further in co-creative manned due to issues being rather complex.

## 1.4.2   *Subcontracting > Cross-Licensing > Cross-Fertilization*

### 1.4.2.1   Linear Subcontracts > Bilateral > Triple or Quadruple Helix

This links also to the previous dimension; in Intellectual Property Rights (IPR) the ownership in the closed innovation approach is clearly defined within the company, which subcontracts. In the open innovation, IPRs are opened for external companies to be used with cross-licensing channels and in other companies IPR is bought in for certain product or process development needs. In clusters, wide cross-licensing is essential to foster common cluster objectives. In Open Innovation 2.0, the opening is between not only organizations, but also goes much further. Companies see themselves as part of ecosystems, and reflect and harvest ideas from communities and end-users who become co-creators of the new products and especially markets. It is proven by several studies that diversity is the key success factor for breakthrough innovations. The people component is essential in the OI2 paradigm as we target towards new markets with new behavioural boundaries enabled by the techno-social development.

### 1.4.3   Solo > Cluster > Ecosystem

Due to complexity in combining a wide variety of technologies in rapidly changing environments, often even the largest companies do not manage the knowledge and experience they need to perform well. Clusters, i.e. groupings of companies in same sector may give flexibility in capacity issues, but it does not necessarily provide the needed knowledge base outside the core business. Ecosystems include also the users in quadruple helix settings. This is extremely important when creating new markets, products, or services, as only in strong interaction we can at early stage see which inventions are growing to innovations. Co-creativity within ecosystems, sharing engagement platforms and environments are extremely important.

### 1.4.4   Linear > Linear, Leaking > Nonlinear Mash-Up

We have seen the change from linear innovation models, which are strictly sequential, and monodisciplinary to the open innovation model presented by Henry Chesbrough. In these approaches, we have anyway the sequential innovation in our minds and give little space to serendipity. Serendipity is essential to embed into the innovation process by project design. That requires agility of both project structures and mindset for the actions. We see projects moving from constraints given by organizational boundaries to something, which we describe as purpose-oriented actions, often happening between organizations towards a common goal.

### 1.4.5   Control > Leadership/Management > Orchestration

The cultural change is a key when moving towards successful implementation of OI2. Ideas emerge from not only the company or its close collaborators but also increasingly from the crowd, i.e. the ecosystem the company is operating in. The challenge for enterprises is how to position themselves in the ecosystem as attractive and fair players, keeping the momentum of the company offering interesting ideas and co-creation structures. This is very much also related to group psychology: Schwarz Universal Values are a good context to this; successful unicorns rarely have the traditional control mentality, more often they are based on strong leadership driving the company idea forward. Most of the well-known fast-growing companies are based on strong leadership rather than shorter term management. Leadership is a good step towards orchestration of innovation. The orchestrator makes the competencies in the palette to play together, gives the talent visibility, and brings his/her own vision to the final play of the orchestra. The orchestrator is a catalyser, an educator, and a visionary. This kind of resource is a rarity in the output of current management schools.

### 1.4.6 Planning > Validation > Experimentation

We are good to make perfect plans—for yesterday! With the complexity and dynamics we are now working on means that instead of planning we can only make qualified guesses. Implementing AI into decision and planning processes may help planning more into detail, but fundamentally we operate in a rapidly changing environment where validation of products in real world is not good enough. Experimenting in real-world settings, in quadruple helix enables us to see at much earlier stage what scales and what fails. Hence, in experimentation the user's role is critical. The Reverse Innovation Pyramid suggests that the users are the source of ideas together with other stakeholders and must be fully involved from the first beginning of the innovation process.

### 1.4.7 Win-Lose > Win-Win > Win-More: Win-More

Closed innovation often targets to better products but in lesser degree to disruptions and entirely new products and services. Open Innovation 2.0 works at its best when creating new. Hence, OI2 moves the enterprise with the ecosystem into entirely new game to markets where there are no competitors or value offerings yet. Win-more in quadruple helix!

### 1.4.8 Box Thinking > Out of the Box > No Boxes

This change is simple and it involves moving from constrained thinking within a box through out of the box thinking to a world to where there are no boxes. This idea in a sense references Piero Formica's idea of creative ignorance and Frans Johansson's De Medici effect where breakthroughs occur at the intersections of disciplines.

### *1.4.9   Single Entity > Single Discipline > Interdisciplinary*

Several studies have proven that diversity correlates strongly with probability of breakthrough innovations; the richer the competencies in the team the better. Cohesion comes from the orchestration and diversity becomes an innovation engine when disciplines, ideas, and stakeholders mingle in a fluid but tolerant manner. Clusters represent single discipline approach, and enrichment both in stakeholder and topical dimension can only happen in ecosystems with fluid interactions, i.e. breaking boxes is not enough, as the new mentality requires openness and curiosity beyond boundaries. Hence, also the role of the orchestration for common purpose raises in importance for successful innovation.

### *1.4.10   Value Chain > Value Network > Value Constellation*

Mental models are changing from something linear to networked to mash-up where serendipity has a possibility. The value chain represents the linear innovation and operational model; networking is bringing more redundancy, more connections between the competencies/entities in the system. However in most cases, networked operational models do not deal optimally in changing conditions where uncertainty and serendipity needs to be one of the innovation design characteristics. Value constellations, where we have a cloud of different competencies and organizations allow using only those, which are needed, in a highly context-sensitive manner. Like we see stellar constellations depending on who we are, where we are, and what is the time, too. The stars remain but the imaginary connections are changing all the time based on the purpose. A similar kind of virtual organizational behaviour is needed in innovation. A good word for a modern organization could be OrganiCzation combining the organization with the organic nature of development over time.

## 1.5   Three Laws of Knowledge Dynamics

Newton's laws of motion are three physical laws that laid the foundation for classical mechanics. Similarly, we believe there are some emerging laws of knowledge dynamics that underpin the shift to OI2. These have been postulated by Amidon et al. (2004) and expanded upon by Andersson et al. (2009). We will discuss these laws briefly.

1. Knowledge multiplies when it is shared
   This *implies* that a knowledge-based economy, which relies more on intellectual capital than environmental or financial capital can lead to a more sustainable way to satisfy human's wants and needs.
2. Value is created when knowledge moves from its point of origin to the point of need or opportunity.
   This means the primary benefit of knowledge lies in action, i.e. when knowledge is actually used.

3. Mutual leverage provides the optimal utilization of resources both tangible and intangible.
   This law asserts that leveraging the collective knowledge and collaboration creates greater wealth and sustainability for us all.

Information and knowledge are two of the least understood phenomena in our modern world but we believe as science develops we will become better able to understand and harness these compelling resources. The three laws of knowledge dynamics underpin and propel the OI2 paradigm forward.

Also in the work of New Club of Paris by Prof. Lin and Prof. Edvinsson the connectivity and dynamic interaction of knowledge is highlighted. On national studies, competitiveness is extremely strongly correlated to the structural intellectual capital, i.e. the interaction processes. Work is now going on to extend the theory also to innovation ecosystems.

## 1.6   The Importance of Innovation

It is easy to see why many people are drawn to technical innovation, as according to the OECD it is the leading contributor to growth in developed countries. In the USA, 75% of US GDP growth since World War 2 has come from technological based innovation, according to the US Department of Commerce. In the last century often, it was a brilliant scientist at a Bell Lab or IBM lab, which drove new inventions and subsequent innovations. Then along came Open Innovation, which is about a systematic process where ideas can pass to and from different organizations and travel on different exploitations vectors for value creation. Open Innovation was based on the idea that not all of the smart people in the world can work for your company or organization and that you have to also look outside the organization for ideas. Procter and Gamble are frequently referenced as a role model for practising open innovation and their 'connect and develop' open innovation strategy has resulted in almost 50% of their new products emanating from ideas and innovations which started outside of the company. Conventional open innovation is now mainstream with even jewellery companies like Swarovski having more than 100 open innovation partners.

Innovation is very easy to talk about, but much harder to do. Innovation is at the core of Europe 2020 vision equally in the USA the importance of innovation was underscored in President Obama's 2013 State of the Union address he stated, 'Innovation just doesn't change our lives, it is how we make a living'. According to the OECD[1] Innovation is the leading contributor to growth in the leading economies of the world. According to the US Department of Commerce, 75% of US growth since the Second World War came from technological innovation.

We often confuse innovation and invention. Innovation is beyond invention and ideation. Innovation is about making things to happen; the offer meeting the needs. Setting that into the modern dynamic societal, technical, and economic environment means that inherently the speed and dynamics has changed; we also need to have a

---

[1] OECD 2010 Innovation report.

new look at the roles of the stakeholders creating the new markets, i.e. the new value propositions based on common values in a different way than previously. Innovation has changed from the traditional science-based linear models to more complex ones involving stakeholders, different disciplines, and serendipity.

According to the OECD, 80% of all value comes from innovation adoption with just 20% from the production of innovation. It seems that there is an innovation paradox with most innovation efforts, investment, and dialogue focused on the front end of innovation. The front end of innovation is of course important but the big yield comes from adoption of the innovations.

## 1.7  Basic Versus Applied Research and Innovation

It is interesting to contrast the investment focus in different geographies on the portfolio of basic and applied research. According to the European High Level Group on key enabling technologies, the USA, Japan, China, and South Korea all focus at least 70% of their total public R&D budget in applied and experimental development activities with the remainder in basic research. In complete contrast, the European Commission prior to Horizon 2020 allocated 77% of the total budget to basic research. Arguably, this is one factor in the relatively sluggish growth of the European region compared to China, Korea, and the USA. The case of Japan is interesting; Prof. Clayton Christensen argued at the 2014 Drucker forum in Vienna that the reason for Japan's sluggish growth was an overwhelming focus on 'efficiency improving' innovations (which actually eliminate jobs) rather than 'market creating' innovations which fuel growth.

We claim that OI2 is extremely powerful especially when creating "new" in the quadruple helix setting, be it markets, services. or products (Fig. 1.3).

According to Erik Brynjolfsson of MIT. if all IT innovation stopped there would still be decades of organizational innovation that could take place with existing technologies. Large research programs such as the EU FP7 have potentially placed undue emphasis on increasing the rate of invention and idea generation while the paradox is that the value from innovation comes from the adoption of innovation.

The discipline of innovation is constantly evolving and now the combination of exponential technologies together with participation of actors from across value chains is creating a new primordial soup, which creates an environment to yield ever more complex and compelling innovations. Indeed, the unit of competition is changing in that it is no longer how good an individual company or organization is but the strength of the ecosystem in which they participate in is often the differentiating factor for great success, mediocrity, or even failure. Witness the decline of once leading mobile phone handset companies like Nokia and Blackberry, and the unprecedented success of the Apple iPhone and various Android-based handsets. A key differentiator has been the strength, incentivizing and nurturing of the ecosystem developing and using the products. Organizations can no longer afford to do it all on their own as innovations are so interconnected and are often composed of

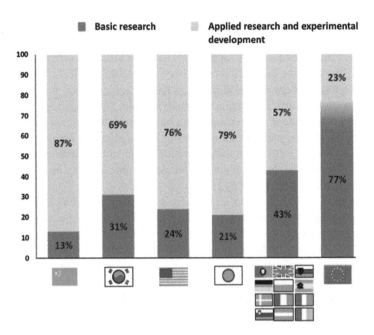

**Fig. 1.3** Strategic focus on applied versus basis research (source OECD)

intelligent combinations of emerging and existing solutions and building blocks. Much of future progress will be driven by collaborative and open innovation—how to execute and govern this is a key question?

## 1.8  OI2: A New Mode of Technical and Societal Innovation and an Emerging Pattern Language

OI2 is a new mode of technical and societal innovation. The notion of a community or ecosystem co-innovating together is central to the new mode of Innovation. Increasingly, we are seeing exemplars of great results from collective ecosystem or community innovation. The Society for Worldwide Interbank Financial Telecommunications (SWIFT) is one of the earliest examples of OI2 at work with the Innovation Value Institute at Maynooth University and the Alcatel Lucent led Green Touch consortium are two more recent examples where a global community innovating together have driven strong results. The metaphor of linear momentum applies well here, being the product of mass by velocity, so the ecosystem with the greatest number of participants and co-innovating the fastest will ultimately likely be the most successful. Implicit within is the recognition of the power of the crowd and the growth of both crowdsourcing and crowdfunding is a leading indicator for the future importance of mass collaboration.

Given the array of opportunities that is available, how can these opportunities be most efficiently and effectively harnessed? Innovation itself is a risky business with high failure rates; however, the application of innovation design patterns can substantially improve the productivity of collective innovation efforts. We in the EU Open Innovation Strategy and Policy group have been studying, practising, and publishing an annual open innovation summary for over 5 years and are attempting to codify this new mode of innovation into a new pattern language, i.e. a series of design patterns. Design patterns are nuggets of knowledge and help us remember insights about design and can be used in combination to help innovate solutions. The goal of this effort is that open innovation can become a discipline practised by many rather than an art mastered by few.

As innovation evolves from an art to a discipline, it is important that there is a common vocabulary for expressing the key concepts and for connecting and relating them together. In this book, we present an elemental set of design patterns with the goal of presenting a body of knowledge to help tackle key tasks in successful ecosystem/platform innovation. We present an elemental pattern language to create a platform for OI2 which can be built upon and extended by others to help improve the predictability, probability of success, and profitability of ecosystem-wide innovation efforts. A pattern language is simply a method of describing good practices or patterns of useful organization within a particular domain.

This new era of co-innovation requires a culture shift with a requirement to move somewhat away from Adam Smith's 'invisible hand' where the self-interest of actors in an economy leads to some common benefits and more to a 'sharing economy' perspective based on a principle of shared value where actors proactively collaborate and innovate based on a common purpose. Having a shared purpose is the foundational pattern of the new mode of innovation whereby shared vision, shared value underpinned by shared values is at the core of successful large-scale innovation. Where efforts are aligned using a compelling shared vision, people's efforts and intellect are harnessed through commitment rather than compliance resulting in strong synergies. Synergy is simply the cooperation or interaction of a number of organization, which results in an effect or impact greater than the sum of the individual efforts, and this is a core goal of the OI2 approach.

# Chapter 2
# Digital Disruption

Digital disruption is all around us and the different possible impacts of digital technology can reinforce each other in a chain reaction where an industry that has existed stably for a century is transformed in less than a decade, for example, the rise of Amazon and the demise of Borders bookstores in the USA. Amazingly today almost 50% of consumer ecommerce transacted in the USA is transacted over the Amazon platform.

A core impact of Digital is democratization, whereby technologies, which were only available to governments and large corporates, are becoming more available and affordable, creating dramatic opportunities for disruption. In the past industry, disruption was possible but required very significant investment, time, and tenacity but today high-expectation entrepreneurs can marry technology, ambition, and smart business models to quickly compete and displace incumbents. Peter Diamandis of the Singularity University says 'when something is digitized it begins to behave like an information technology' and thus growth and development develop exponentially and at accelerated time rates.

## 2.1 Pathways to the Digital Revolution

There are a series of pathways which are emerging for digital disruption with a three phase shift from analogue or physical to digital, from single function to integrated multifunction, and then from single system to systems of systems as depicted in the following diagram. Thus, almost in the blink of an eye something that was once expensive and physical becomes an App and costs a fraction of the previous instantiation of something. The migration from an analogue camera through a digital camera to a camera on a phone is an example of this. Polaroid and Kodak both invented in the digital camera but clung to their cash cow film businesses and completely missed the digital camera wave (Fig. 2.1).

© Springer International Publishing Switzerland 2018
M. Curley, B. Salmelin, *Open Innovation 2.0*, Innovation, Technology, and Knowledge Management, DOI 10.1007/978-3-319-62878-3_2

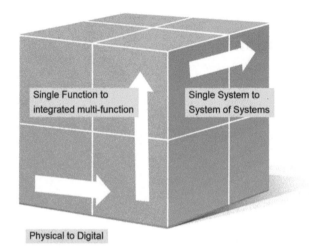

**Fig. 2.1** Pathways to digital disruption

**Fig. 2.2** Six patterns of digital disruption

However then the camera companies, which had invested in developing exquisite digital cameras, were bypassed by the development and integration of high-resolution quality onto mobile phones. This then became the catalyst for the creation of an eco-system of digital propositions. Within a year of Kodak entering Chap. 11 in 2012 (In 1996, Kodak had almost 100,000 employees and a market capitalization of almost $30 billion), Instagram with just 13 employees was acquired for $1 billion by Facebook.

There are many ways for digital disruption to occur and next we will discuss six of the more common emerging patterns for the disruptive impact of digital technology (Fig. 2.2).

## 2.2  Disintermediation

Digital creates the opportunity to eliminate intermediaries from a supply chain between manufacturer and service provider to a customer. The intermediation effect can change the basis of competition in an industry, often reducing costs and providing more choice to customers. Digital can dramatically increase market transparency, where much more information is known about products, services, location, and price by many more compared to a non-digitally enabled industry or market. The increased transparency can make the industry or market overall more efficient.

The term disintermediation was first used in the context of banking and the US Government Regulation Q, which limited the interest rate paid on interest bearing accounts that were protected by the Federal Deposit Insurance Scheme. In response, customers avoided the intermediation of banks by investing directly in securities. While that instance of disintermediation was driven by regulation, this wave of banking disintermediation will be driven by both technology and regulation. In the UK, Fintech start-ups such as Revolut aim to disintermediate by changing the basis of competition offering customers the possibility of customers using the interbank rate directly for foreign currency transactions and enabling the set-up of current accounts in 60 s. N26 in Germany offer similar offerings and both offer contactless Mastercard for low-cost debit transactions.

In Europe, the Payment Services Directive 2 is requiring banks to open up customer account information, by consent, using Application Programming Interfaces (APIs) which will disrupt and change the basis of competition. In this particular case, the regulatory change is only possible because the digital technology makes it possible.

## 2.3  Distribution

Digital technology disrupts through dramatically improving access to customers and dramatically reducing the cost of distribution of services, products, and solutions. As the information intensity of products increases, they are increasingly more easily distributed and the cost of distribution is dramatically reduced, for example, distributing a song via an MP3 file is much cheaper and much faster than distributing it through physical media such as a CD. Similarly for a bank, the cost of a mobile transaction is 95% less than a branch transaction and can be accomplished much faster online rather than being present in person in a branch.

## 2.4  Democratization

Digital enables the democratization of services, making them more available and more affordable than would otherwise be the case. Today, cloud computing makes computing services, on a buy as you go basis available to start-ups all around the world, allowing the same access and scale that previously was the domain of large corporates and requiring substantial upfront capital investment. According to Microsoft, the cost of acquiring a cloud service from their Azure Cloud platform is less than one fortieth of buying, installing, and operating the server yourself. Cloud computing operators such as Microsoft and Amazon Web Services offer economies of scale and scope, which are enabling democratization of computing services.

Case Study: Facebook Open Compute Project.

The same can be also seen in opening up not only intangible components but also hardware. Arduino is a very good example on how hardware has been made afford-able to masses in the spirit of Open Innovation.

The Open Compute Project (OCP) is reimagining hardware, making it more effi-cient, flexible, and scalable. Founded by Facebook, a community of technology leaders are working together to unlock the black box of proprietary IT infrastructure to achieve greater choice, cost savings, and customization.

Often with these kinds of initiatives, there is a champion who has the vision and perseverance to sell the idea, recruit the community partners, and to make it happen. Frank Frankovsky, the then director of Data Centers at Facebook was the visionary, who when faced with the escalating demand, cost, and energy consumption turned to an open innovation approach to meet the challenges. Facebook designed the world's most energy-efficient data centre in Prineville, Oregon, and in 2011 shared its designs with the public and along with Rackspace, Intel, and others created the Open Compute Project foundation. The idea was that the community would create the same type of collaboration and creativity that the open source software movement created. The OCP foundation believed that openly sharing ideas, specifications, and other intel-lectual property were pivotal to maximizing innovation and reducing complexity.

Demonstrating that open innovation works, Facebook estimates it saved $1.2 billion through smarter, more energy efficient and quicker hardware designs in a 3-year period. At the Open Compute Summit in 2014, Facebook CEO Mark Zuckerberg and VP of Engineering Jay Parikh highlighted the cost savings and other benefits.

## 2.5  Dematerialization

Digital dematerialization can take place in many forms, for example, providing real-time information of what new fashions are selling and which aren't in stores cou-pled with just-in-time manufacturing and distribution can avoid inventory right offs and allocation of scare production capacity to 'hot' products. Equally, 3D manufac-turing can help eliminate expensive distribution costs when designs for parts can be sent electronically and printed at the location where the part is required. A compel-ling example of dematerialization is where separate bulky physical products are integrated into a single product, for example, cameras, GPS navigation systems, CD players, and even physical maps are now integrated into a smart phone which has become a form of Digital Swiss army knife. This dematerialization is also calling for new company structures like virtual enterprises or fractal factories where agility combines with flexible capacity.

## 2.6   Demonetization

Digital is simultaneously enabling digital currencies such as Bitcoin and taking money out of the implementation and adoption equation. On the one hand, there are new App industries worth billions of dollars with hundreds of thousands of developers but on the other hand, silicon, software, and network economics mean the cost of developing and using solutions and services are decreasing continuously, often approaching to or actually being free. The cost of DNA sequencing has dropped even faster than the cost of computing driven by Moore's law with the cost of sequencing an individual's genome in 2015 roughly costing $1000 compared to the costs of over a million a decade previously. On a smartphone, there are many apps available for download, which all provide utility used for almost zero cost. In parallel, some luminaries such as Don Tapscott have hailed block chain as one of the most significant innovations in computer science and many speculate that a myriad of financial technologies innovations will be enabled through the distributed ledger capability (DLT) it enables through the secure peer-to-peer distributed database block chain mechanism. DLT will be one of the key elements in trusted distributed industry commons for new business model development.

## 2.7   Deceptive Displacement

A disruptive technology is often initially inferior to an incumbent solution and is not perceived as a threat, however due to the exponential nature of Digital, something that starts as insignificant due to exponential growth can become very big very fast. This so-called big bang disruption can happen very fast. Compare the meteoric rise of WhatsApp to an operator like Vodafone where WhatsApp grew dramatically and the pendulum of value capture swung towards the Digital world. Something can start small but because of exponential growth quickly outpaces the linear growth of more traditional products and services. Occasionally, the so-called big bang disruption also happens—today an Internet meme can be diffused around the world in a matter of days. Angry Birds was downloaded over a million times within the 24 h of the Android release and subsequently there were over two hundred million downloads within 7 months of the first release.

## 2.8   Attributes of Digital

What makes Digital so potent compared to other innovation ingredients? First, it is unique in that it is both an innovation infrastructure and an ingredient and out of the innovation process. Digital connectivity dramatically lowers collaborative friction and dramatically extends the access to collaborators for and adopters of innovation.

### 2.8.1   Malleable

Unlike physical infrastructure like buildings, railways, and generators, which fuelled previous industrial revolutions, Digital infrastructure is much more malleable and flexible. With virtualization technologies, compute loads can be moved seamlessly across datacentres and data borders. This flexibility allows computing to be moved close to the loads or workloads to be sent to the most appropriate datacentre (Fig. 2.3). The same applies also in e.g. 3D manufacturing where the actual production can be very local whilst the databases for the designs are globally accessible.

### 2.8.2   Programmable

Programmable means being capable of being programmed for automated operation or computer processing. Increasingly, we will hear the term 'software defined' appended ahead of nouns like cities or electricity grids. Take for example a city, which increasingly will be digitized and connected and the opportunity will exist to program and run the city based on different logic or indeed business models because of the ubiquitous nature of digital. Once there is a software-defined city one could program the city's traffic system to optimize on lowest commute times for citizens or to also optimize across minimizing fuel consumption and environmental impact. The smart phone is the Swiss army knife of the digital era and today's iPhone has more computing power than the entire compute power of the Apollo missions which sent a man to the moon and back.

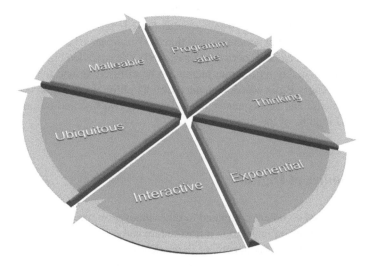

**Fig. 2.3**  Potent attributes of digital

### 2.8.3 Thinking

With the emergence of machine learning and artificial intelligence, computers are becoming more and more thinking machines, with the ability to reason and provide substantial decision support. Artificial intelligence is the increasing ability for computers to perform tasks that normally require human intelligence such as translation between languages, speech recognition, visual perception, and decision-making. In AI, a computer mimics cognitive functions that we associate with human thinking. Machine learning is a subset of AI that enables a computer to learn without being explicitly programmed. Machine learning uses analysis of data to detect patterns in data and adjust program actions accordingly.

### 2.8.4 Exponential

Driven by the nature of silicon, network, software, and storage economics, digital technologies continue to improve in the performance/cost ratios year after year. There are a variety of the so-called Laws which describe the advance of capabilities, Moore's law of processing power, Gilder's law of optical fibre transmission rate improvements, etc.—the net result is that each year we can do a lot more for less and Digital then becomes a catalyst and a raw material for further digital transformation.

### 2.8.5 Interactive

Increasingly, we have seen how computers have become much more interactive with the ability to interact with humans in a conversational way. Whether it be Apple's Siri or Amazon's Alexa the ability to interact with a computer, almost in a human-like way is a game changer. User experience and user interfacing continue to improve and future breakthroughs in haptic interfaces will better enable humans to interact through bodily sensations and movements. Gesture-based controls such as Microsoft's Kinect will continue to improve and become more and more pervasive.

### 2.8.6 Ubiquitous

Computers and data are becoming ubiquitous. According to Ericsson, there will be over six billion smartphones in the world by 2020 and according to Cisco, there will be over 50 billion connected things by 2050. As computing power and networks continue to grow, the planet will be immersed in a giant web of technology, which already gives us information at our fingertips but increasingly will give us actuation at our fingertips, where will we be able to remotely control objects at will.

Together these properties make digital one of the most potent resources available to us to and one, which is a key catalyst and enabler for a better future. This future we call Sustainable Intelligent Living.

## 2.9  IT-CMF (Information Technology Capability Maturity Framework)

Digital, as we have seen is a very potent set of technologies but in parallel a systematic approach is needed to define, design, deliver, and operate systems based on digital and information technologies. At the Innovation Value Institute, we have created a pattern language for chief information officers and other executives to systematically manage and improve digital and information systems. A capability maturity framework, which in turn is supported by thousands of best practices captured and created by a global team of practitioners, and researchers who used an OI2 approach to build the framework underpin this pattern language. The framework recognizes that it is not just the technology that needs to be managed but that comprehensive approaches to domains such as Enterprise Architecture, IT and Digital Governance, Risk Management, and Supply Demand Management need to be in place as well as those for solutions delivery and services provisioning. Here, we present a short summary of the IT-CMF. A more detailed elaboration on the IT-CMF is beyond the scope of this book but can be found in the recently released IT-CMF Body of Knowledge book (Curley et al. 2016).

### 2.9.1  IT-CMF Rationale and Summary

As the move to Digital becomes a mega trend, IT is moving from the backroom to the boardroom. It is not sufficient to put on digital lipstick but digital transformation must be driven by simultaneous digital and IT transformation. Organizations, both public and private, are constantly challenged to be increasingly more agile, innovative, and value adding. CIOs are uniquely well positioned to seize this opportunity and adopt the role of business transformation partner, helping their organizations to grow and prosper with innovative, IT-enabled products, services, and processes. To succeed in this, however, the IT function needs to manage an array of interdependent but distinct disciplines.

In response to this need, the Innovation Value Institute, a cross-industry international consortium, developed the IT Capability Maturity Framework (IT-CMF). The IT-CMF represents a suite of capabilities, see below that help improve the management of IT to deliver higher levels of agility, innovation, and value creation (Fig. 2.4).

The IT-CMF consists of a set of integrated and connected critical capabilities which when managed and improved together drive agility, innovation, and value. The sum of critical capabilities form a periodic table of the atomic level capabilities

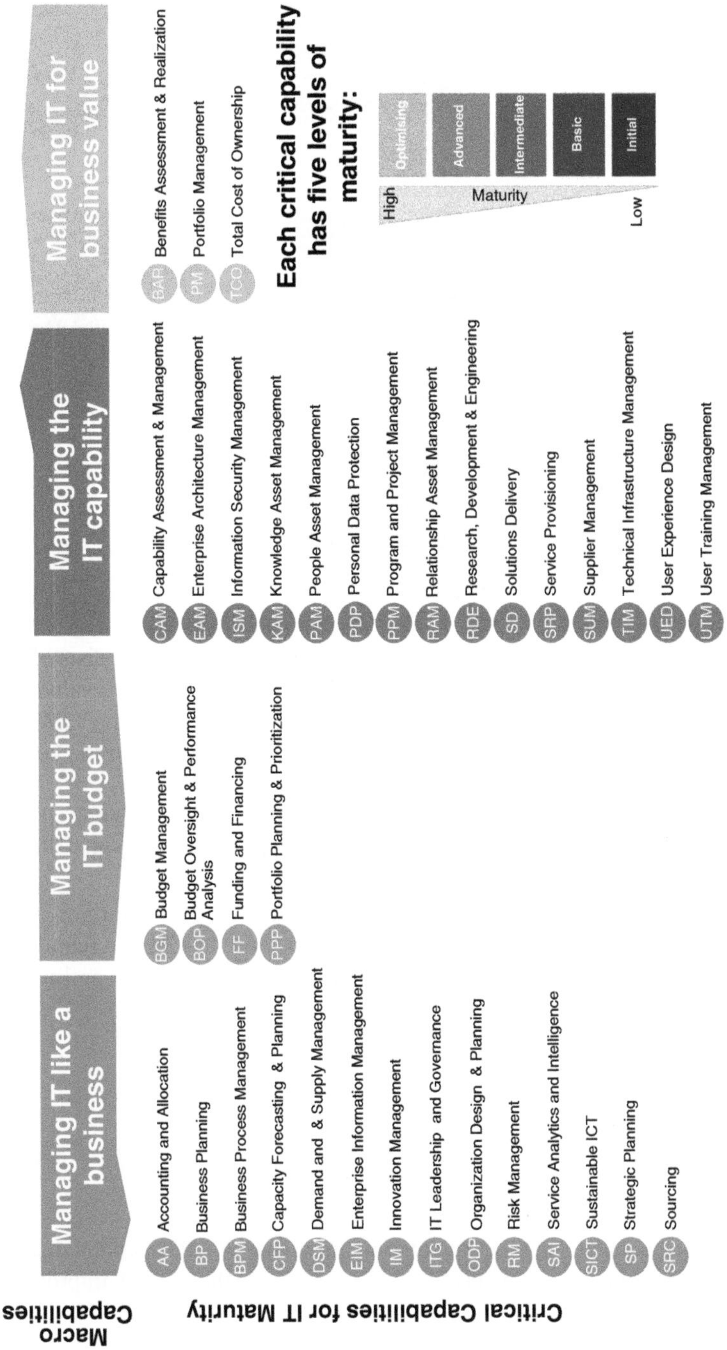

**Fig. 2.4** IT Capability Maturity Framework—critical capabilities

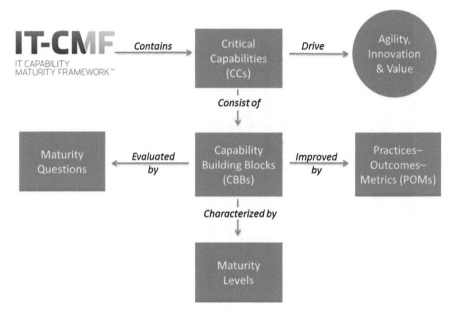

**Fig. 2.5**  IT-CMF artifact type architecture

necessary to achieve world-class IT and business. Each capability consists of capability building blocks which are characterized by maturity levels, evaluated by maturity questions, and improved by practices-outcomes and metrics (POMs). The following artefact architecture figure shows the relationship between the key artefacts of the IT-CMF (Fig. 2.5).

Each critical capability of IT-CMF consists of capability building blocks (CBBs), which are characterized by maturity levels, are evaluated by maturity questions and are improved by practices-outcomes-metrics (POMs).

Business and IT professionals seeking to harness the full potential of digital and information technology in their organizations can use the IT-CMF to demonstrate leadership in adopting better approaches to managing technology for agility, innovation, and value impact.

## 2.9.2  Framework Description

**The IT-CMF is:**

- An integrated management toolkit covering more than 30 management disciples, with organizational maturity profiles, assessment methods, and improvement roadmaps for each.

- A coherent set of concepts and principles, expressed in business management terms that can be used to guide discussions on setting goals and evaluating performance.
- A unifying (or umbrella) framework that complements other, domain-specific frameworks already in use in the organization, helping to resolve conflicts between them and filling gaps in their coverage.
- Industry/sector and vendor independent. IT-CMF can be used in any organizational context to guide performance improvement.
- A rigorously developed approach, underpinned by the principles of Open Innovation and guided by the Design Science Research methodology, synthesizing leading academic research with industry practitioner expertise.

**Considerations when using IT-CMF in your organization**

- IT-CMF should be adopted at a senior management level, not just at the front-line staff/practitioner level to realize the framework's full advantages.
- The framework delivers value through change. Committing to the required organizational change will help ensure expected outcomes are achievable.
- Appropriate capability selection and setting of maturity targets benefits from an appreciation of IT-CMF's critical capabilities, as well as a blended view across an organization's business strategy, IT posture, and industry context.

# Chapter 3
# Sustainable Intelligent Living

*'Technology is a resource liberating force'*

Peter Diamandis

## 3.1 Sustainable Innovation

Sustainable innovation can have several meanings and all of them are desirable. Firstly, innovations that result in better more efficient use of resources and secondly innovations that have longevity. We define innovation (Baldwin and Curley 2007) as the adoption and creation of something new, which create value for the entities that adopt and create/deliver the innovation. The business proposition for a particular innovation is only sustainable if both the creating and receiving entities achieve value more than the cost of creation and delivery and the cost of adoption. Increasingly, we are seeing that the unit of competition is moving to the ecosystem (Curley 2015).

Sustainable innovation is also full of disruptions requiring innovators to have courage to seek the unexpected. OI2 approach creates a safety net for innovation.

The opportunity exists to make a significant improvement to collective sustainable innovation progress by simultaneously more aggressively leveraging information technology and adopting a new emergent innovation paradigm Open Innovation 2.0 (OI2). A key sustainable development strategy is to decouple growth from resource consumption and environmental impacts and in this context; the remarkable productivity/energy efficiency of information technology driven by Moore's law can provide the basis of a viable innovation pathway towards sustainability. While the semiconductor industry has played its part by dramatically improving the energy efficiency of computing, a much bigger opportunity exists by leveraging IT to automate, dematerialize, and substitute for other natural resources in value chains. The adoption of the UN Sustainable Development goals (UN 2015) in September 2015 and the COP21 meeting in Paris in December 2015, COP 21 provide strong platforms to catalyse efforts to create a sustainable future. The emergent OI2 paradigm and associated design patterns can be used to align, amplify, and accelerate innovation efforts to deliver more sustainable solutions for societal scale systems.

© Springer International Publishing Switzerland 2018
M. Curley, B. Salmelin, *Open Innovation 2.0*, Innovation, Technology, and Knowledge Management, DOI 10.1007/978-3-319-62878-3_3

## 3.2   Sustainable Intelligent Living

OI2 is not just a new methodology for innovating, but is also about innovating towards a purpose. The overall vision of OI2 is about enabling sustainable intelligent living by helping co-innovate solutions that simultaneously improve economic growth and human well-being but also reduce per capita consumption of natural resources and reduce environmental impact. OI2 is very aligned with the thinking associated with the circular economy and indeed goes beyond that to what Stahel (2010) calls the performance economy. Thus, the four characteristics of OI2 outcomes are

- Improved economic or service benefits
- Improved standard of living or improved well-being
- Reduced per capital consumption of natural resources
- Reduced environmental impact
  The following figure shows the desirable situation whereby resource and impact decoupling can be achieved (Fig. 3.1).

OI2 deliberately seeks innovations which not only improve economic growth but also human well-being while simultaneously reducing resource usage and reducing environmental impact. OI2 is not just about the 'how' of Innovation but the 'what'. Peter Drucker often said that the problem for managers is often not 'how to do' but 'what to do'. Sustainable Intelligent Living provides a focal point for dramatically improving the results and holistic benefits from coordinated innovation efforts. Innovation capacity is most powerful when it is mobilized in the context of a compelling shared vision. OI2 proposes 'Sustainable Intelligent Living' as an overarching vision where innovation efforts are focused on delivering intelligent innovations which when using the power of information and other technologies, result in new products and services that are people centric and are better than previous offerings, easier to use, and very importantly are more resource efficient than previous generations of products. We have all seen how IT has transformed the music, book, and

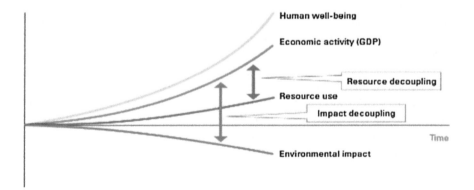

**Fig. 3.1**   Sustainable intelligent living, source UNEP International resource panel

banking industry led by companies such as Napster, Apple, and Amazon, for example. Imagine the possibilities if we could deliver similar transformations in our cities, healthcare, transformation, and energy systems. While these transformations are much more complex, the OI2 paradigm and methodology are targeted exactly at enabling these kinds of transformations.

### 3.2.1 Sharing Economy and Collaborative Consumption

The sharing economy is transforming many of the traditional industry and service sectors. Good examples of this are Uber and Airbnb where entirely new business models challenge the traditional ones, and even legislation and the regulators. The new business models would not have been possible without sharing platforms and power of crowds to provide and consume the actual services. A sharing economy is enabled by digital technologies, empowering different entities through information that allows sharing, redistribution, and reuse of excess capacity of products and services. A core concept that underpins the sharing economy is that when information about a product or service is shared, the value of these goods may increase, and this is an exemplar of the laws of knowledge dynamics at work.

Collaborative consumption is a phenomenon in which entities share access to services or products rather having individual ownership. A preference for access to services and products rather than owning products is becoming more mainstream with the WEF reporting that more than 110 million people in the USA already participating in the sharing economy using services like Uber and Airbnb. This shift towards access-based consumption models can be a key driver towards more sustainable intelligent living and a disruptor for more traditional business models. This is complemented by an ongoing shift from a physical to a digital world. As evidence of the shift Airbnb chief marketing officer, Jonathan Mildenhall, describes Airbnb as the first community-driven super brand.

In collaborative consumption, the platform approach is evident. The division of the 'fair share' of the collaborative consumption to the stakeholders involved in the value creation process is critical.

## 3.3 Sustainable Development

Sustainable development as a concept can be defined as 'development that meets the needs of the present without compromising the ability of future generations to meet their own needs' (World Commission on Environment and Development 1987). The concept of 'doing more with less' is a concise encapsulation of this concept. Many reports warn of the consequences of ignoring the need for a new sustainability paradigm. The global footprint network has stated that if current consumption trends continue we would need two world's worth of resources to support us by 2050.

EF Schumacher, the German economist who worked for the UK National Coal Board, was able to capture the essence of sustainability in his phrase "small is beautiful" and this is a good lens through which to consider Moore's law.

The impact of innovation becomes very profound when the intersection of the emergence of technology and grand societal challenges occur. The fast pace of development of emerging technologies offer opportunities and solutions for challenges in areas such as agriculture, cities, and health and in parallel offer the opportunity to create sustainable business and profit opportunities. Disruption on social models will challenge the smart cities vision as well; changing nature of mobility (as service), new ways of work, and new roles of public sector to provide services increasingly location independent but cost-effectively will change the shape of future cities dramatically. We are in a transition far beyond smart cities being only cities full of sensors and data.

## 3.4  Digital/Moore's Law and Resource Decoupling

Moore's law is one of the few modern business phenomena, which actually support the sustainability paradigm. The computer industry has certainly played a part. An ACEEE[1] study has shown computing energy efficiency improved by more than 2.8 million percent in the 30 years from 1978 to 2008 while other industries such as lighting, automotive, and agriculture achieved orders of magnitude less energy improvement. The introduction of multi-core technology in microprocessors indeed has accelerated this progress even further and the continuous shrinking of geometries has allowed the semiconductor industry to achieve a 'less is more' outcome where each successive generation of product is smaller, more powerful, and more energy efficient than the previous generation. As a consequence of this routinely one new server today replaces ten older servers and consumes a fraction of the rack space in a datacentre.

Gartner estimates that the IT industry contributes roughly 2% of the world's energy greenhouse gases but it is increasingly recognized that the big opportunity in the future is to substitute information and technology for other resources across the economy and indeed the broader society. Indeed, another ACEEE report identified that some of the most energy-intensive industries such as transportation were amongst the lowest in IT intensity indicating strong potential for significant energy savings through more deployment of IT. The Intel SAP co-Lab in Belfast recently showed that through smart scheduling algorithms that approximately 100,000 new electric vehicles (EVs) could be introduced onto the island of Ireland without any significant increase in electricity generation capacity. Imagine what also might be possible when each of the EVs acts also as an energy storage device in a smart grid.

Self-driving cars might also be a game changer in urban areas as it may reduce the need of owning a personal car. Proper simulations for policymaking are needed

---

[1] 'A Smarter Shade of Green', ACEEE Report for the Technology CEO Council, 2008.

when elaborating the impact of new scenarios, especially when we look at the impact of technology to change of work and living, directly influencing the city planning, and therefore also the mobility needs.

While acknowledging the ongoing contribution of future ultra-low power IT devices and green software to directly improving energy efficiency, the broader impact of ICT to future sustainability will be through mechanisms such as

- Energy and resource efficiency achieved through sophisticated sensing, analytics, and actuation
- Dematerializing through substituting ICT resources for other inputs (e.g. labour, materials) in value networks and indeed broader societal activities
- Enabling servitization, a shift from products to services, where the goal shifts from to attempting to maximize product consumption to optimizing service whilst optimizing product longevity and resource usage

The combined effects of such mechanisms can help achieve 'resource decoupling', whereby economic growth and quality of life can be improved, while simultaneously reducing resource consumption and environment degradation. One glimpse of the impact of such mechanisms is the finding in another ACEEE report which estimated that for every 1 kWh of energy consumed in a US datacentre, an average of 10 kWh were saved across the broader US economy. Logistics companies such as UPS have claimed fuel efficiencies of approximately 20% through using smart scheduling software to maximize the number of right hand turns delivery drivers are able to make on their daily delivery runs. The Innovation Value Institute in Ireland have developed and defined a Sustainable Information and Communications Maturity Curve and Assessment instrument to help organizations assess and improve the impact of applying ICT for sustainability.

Better energy management with distributed electricity sources has a clear transformative role in the markets. Lowered cost of photovoltaics leads to stronger local economy even on household level. This will have a longer term impact on the power grid management and energy storage, which interlinks the energy cluster strongly to other ones, e.g. the mobility one. This is a good example where sustainable development leads to multidisciplinary ecosystems with citizens as important players as well.

These mechanisms will and are already creating the potential to accelerate a transition to a new sustainability paradigm. We are seeing a virtuous circle where the cost of information is falling and the availability of information is dramatically increasing as wireless, LTE, and indeed fibre infrastructure continues to be built out. Almost unnoticeably the mobile Internet has emerged and we can envisage a world where if something has electrical power (which may be scavenged from the ambient environment), it will compute and communicate. This opens up the possibilities and power of ambient intelligence.

### 3.4.1   Substitution, Automation, Dematerialization

Year on year as information becomes cheaper, more of it can be substituted for other inputs such as natural resources, labour, and capital in both business and societal activity. eBook readers such as Amazon Kindle are a good but not well-reported example of enabling sustainability. Fewer natural resources are consumed through production and distribution of eBooks than physical books and also readers have much more choice of what they want to consume and the books are cheaper. This is a good example of the sustainability paradigm at work almost unnoticeably. It is also a brilliant example of what McInerney and White describe in their book 'Future Wealth'—'our ability to substitute information at will for every other resource, including capital and labor, is altering the economy (and indeed society) in a fundamental way'. Skype is another example of a disruptive innovation with a similar 'sustainability' effect allowing relationships to be better sustained, business to be more efficient all at a lower cost and with more efficient use of resources. Furthermore, professional video conferencing services can create remote meeting experiences, which are as good as if not better than physically, meeting saving time, money, and resources while improving productivity.

These technical trends will also shape the future job infrastructures. We see increasingly flexible work centres being established in municipalities, reducing heavily the mobility needs for work. This is strongly the case in ICT and knowledge intense jobs, e.g. engineering. The co-working spaces are going beyond the telecentres, as they are combining the co-work space with prototyping facilities and maker spaces too. These centres bring the innovation and prototyping environments better accessible to the local community level as well.

## 3.5   Servitization

The shift from product to service, sometimes called servicizing, to meet dual goals of improved profitability and sustainability benefit is often enabled by ICT. In servitization, the focus is not on maximizing the consumption of products, but on the services the products deliver with one output being improved resource efficiency. The adoption of cloud computing and software as a service are contemporary examples of this. Rolls-Royce with their 'power by the hour' program whereby they sell hours of flying rather than aircraft engines, all enabled by remote telematics was a pioneering example of this concept.

However, energy is also getting personal. Increasingly policy and legislation is requiring that consumers have better information on their electricity consumption. A study on smart metering in Ireland showed that more than 80% of consumers who had better information on their consumption made a change to their behaviour causing a reduction in demand. Intel labs developed personal office energy management software to create new infrastructure using personal and participative computing to create much richer building energy management information while helping users to monitor and change the comfort and energy consumption of their own environments.

## 3.6   Digital and Sustainability

Moore's law is one of the few modern business phenomena, which actually supports the sustainability paradigm, and the doubling in transistor density every other year delivered at less or equal cost with increasing energy efficiency has continued for several decades. Information technology has hugely outperformed all other industries in improving output per unit of energy input over a 30-year period (Technology CEO Council 2008). Whereas, for example, the automobile industry demonstrated a 40% energy efficiency improvement over a 30-year period, the energy efficiency improvement of IT over the same 30-year period was a staggering 2,857,000%. While this performance is impressive, the real impact is achieved when IT is applied to make other industries and societal systems more efficient. The American Council for an Energy Efficient Economy found that 'for every kilowatt-hour of electricity used by IT, the US economy increased its overall savings by a factor of about 10' (Laitner and Erhardt-Martinez 2008). IT helps improve energy efficiency by automating and optimizing other energy using systems and by substituting, IT-related services for other services and products across the broader economy. As the Internet of Things emerges, smart connected, collaborating things will exponentially increase the opportunities for energy, resource, and performance optimization across systems.

## 3.7   Plan C: Decoupling Natural Resource Use and Environmental Impacts from Economic Growth

While it is clear that we must collectively embrace the sustainability paradigm, there is collective inertia in developing and adopting solutions, which might take us there. The UN Sustainable Development goal number 12 is focused on responsible production and consumption; however, there is a perception that we must all sacrifice convenience and functionality to adopt these new green solutions. However, the use of IT enables the possibility to create solutions, which are not only more resource and ecologically efficient but also provide better functionality and convenience. These kinds of solutions exhibit the characteristics of *Plan C* type solutions (Curley 2013). Plan C type solutions exhibit a number of characteristics—they do something better or different than is done today, are more energy or resource efficient and offer more convenience than the existing method or product. Digital books, banking, and music are all examples of Plan C patterns, offering services, which are generally better, cheaper, and more convenient than the incumbent physical goods and services that they are displacing.

Often these kinds of products/services benefit from *network effects* and thus once they hit critical mass they are widely adopted quickly bringing both user benefit and positive ecological impact quickly. Specifically advocating for the design, development, and adoption of goods and services, which exhibit these kinds of characteristics, will not only drive energy and resource efficiency but also improve quality of service and life.

Industries that existed for decades are being transformed almost overnight through the innovative use of IT—witness the music industry, catalysed by Apple and the book industry catalysed and led by Amazon. These two examples also illuminate a significant pattern that as IT enables Schumpeter's 'creative destruction' to transform industries, these industries are made more energy efficient but simultaneously create opportunities for better service, better scaling, and better access. These parallel benefits significantly lower the barriers to adoption of new efficient innovations. There are many other industries ready for disruption with the electricity grid being a prime candidate. In particular, the adoption of the EU Energy Union initiative highlights the opportunity to create a virtuous circle of innovation using ICT to deliver better use of distributed renewables, improved energy efficiency, and more climate-friendly energy supply.

### 3.7.1  Cities as a Focal Point for Sustainable Intelligent Living

More than 50% of the world's population live in cities, and cities consume more than 75% of the world's energy as well as contributing about 80% of greenhouse gases. Cities are at the confluence of three mega trends—digital transformations, sustainability, and mass collaboration that we have discussed previously.

In London, UK, together with Imperial College London and University College London, Intel launched a Collaborative Research Institute for Sustainable and Connected Cities. Together the stakeholders saw the opportunity to collectively research and create an evolutionary leap for cities in terms of resource efficiency, new services, and ease of living. Using a quadruple helix innovation approach involving Government, Industry, Academia, and City Dwellers the stakeholders were trying to catalyse and drive a step change in the outcomes of innovation for energy efficiency and sustainability in cities. As part of the research, big data from a distributed ambient intelligence platform is providing the opportunity to perform real-time optimizations and city visualizations as well as enabling a completely new set of services. The mayor of London created the Smart London Board including leading academics, businesses, and entrepreneurs—to advise on how London can put digital technology at the heart of making the capital an even better place to live, work, and invest.

The city of Dublin led by the then Lord Mayor Naoise O'Muiri and Director of Research and Internationalization Peter Finnegan also launched a sustainable cities initiative and has for a number of years regularly convened a quadruple helix innovation line up of experts to advise and help implement solutions. Dublin has also launched a series of SBIR (Small business innovation research) grants to encourage organizations to submit innovation proposals against gnarly city challenges.

ICT is also the critical enabler for the co-working spaces, which reduces the pressure for the growth of cities and mobility of work. It remains to be seen whether the trend of growth of megacities will be reversed, and when. Megacities will likely also develop smaller communities within reducing the need for mobility internally. ICT will have a crucial role in the internal reorganization and management of these megacities too.

**Case Study: Open Innovation and Dublin—A Journey into Future**
**To leave a mark upon the ground I have walked**
  **To make a difference in the lives I have touched**
  *The key to handling future unknowns lies in the courage to risk innovation.* The success in achieving innovative solutions and driving change rests in the openness to seek out, share ownership with and embrace partners, even competitors, who become co-creators. People, as individuals and as decision-makers and shapers in organizations are more important than the tools of technology in driving and delivering effective and relevant innovation.

In Dublin we started, within the context of developing a 10-year strategy for economic, social, and cultural development, to identify the impact and opportunities that digital technologies would have on the city experience. *'Dublin-City Of Possibilities'* emerged in 2001 as a Policy framework, which emphasized openness, innovation, and collaboration as key elements to anticipating and creating change in an increasingly connected and digital world. At that time, our focus was on the Knowledge Society. Two key events that prove the power of synchronicity conspired to create the initial theoretical framework and connections that led to Open Innovation 2.0 become the guiding principles driving innovation initiatives in Dublin. The first was our participation in an EU project called **Intelligent Cities for the Next Generation** (ICCING); the second was the first encounter with Prof. Martin Curley when he spoke at our International ICCING Conference.

ICCING was a collaboration between three cities Local Government structures, Business (Multinational and SME), and Higher Education research. It became a model for broader networks of collaboration around the innovation agenda, both locally and globally. The realization that a key component and potential in the digital future was the active participation of the citizen led to the fourth strand of the helix becoming a partner to the initiative.

Working through Prof. Curley, the research arm of INTEL became a key partner in various projects around challenges faced by Dublin. This connection also led to the *Innovation Value Institute* (IVI) taking on the task of adapting an existing impact assessment model and developing the concept of a digital impact scorecard for a city.

Many steps followed that culminated in co-hosting with **Intel and the EU Commission**, under the banner of the **OISPG**, the first **Innovation 2.0 conference and Luminary Awards** in Dublin during the Irish Presidency in 2013. Along the way, we explored through **Creative D** the collaborative power of the Arts and Design to grow economic development through interaction with business and digital technologies; we developed an open data initiative that became to be known as **Dublinked**; expanded our knowledge of what was happening through the Dublin Dashboard project; and most importantly worked with and through business, community, and public sector partners to launch in 2013 Dublin's first ever **Digital Masterplan** (www.digitaldublin. ie). The Masterplan was a roadmap around the innovation mindset and actions that city stakeholders needed to take to make Dublin, not a Smart City, but rather a Creative City, which harnessed the power of digital technology to release the creative energy of its people, to enable the application of innovative actions that connected people and improved knowledge, place, and opportunity.

Open Innovation 2.0 provided the theoretical and conceptual framework that captured and expressed what we had been doing as we had made our journey to this point. The principles underpinning Open Innovation 2.0 found echo in the way in which we had worked on the Digital Masterplan.

The strategic vision and perspective was both local and global. The newly formed International Office adopted the digital agenda as a key strand in building networks of collaboration with key innovative cities from the Americas, through Europe to Asia. The sharing and openness between the city administrations, the business communities, the academic institutions, and the citizens reached beyond the individual city and created a global collective energy and direction.

At the heart of the endeavour lay the importance of addressing the needs of people. It focused on building, through the interaction of people, place, and digital technologies, solutions to the challenges that face all humanity, challenges of sustainability, expression, connectivity, belonging, well-being, and of cultural and economic achievement.

The journey continues within a fast moving and ever changing landscape through the work of the **Smart Dublin team** (www.smartdublin.ie). The principles and practice of Open Innovation 2.0 underpin that work and that journey.

Although the descriptive language has changed from Knowledge Cities, to Intelligent Cities, to Smart Cities, the underlying truths remain the same. We live in a changed and ever fast changing world. The speed and nature of that change is driven by the power and reach of digital technologies. Addressing the impact of those changes requires an openness to innovate collectively, reaching beyond the closed walls of institutions and business, beyond the boundaries of cities and communities. It requires the practical ability to collaborate and co-operate within a competitive world. It requires the capacity to think beyond the present and the local; to think, plan, and dream strategically for the future and the global. It requires the realization that the era of 'use and throw away' must be replaced by the building of a more sustainable, equal, and engaged society and world.

*Peter Finnegan,*
*Director of Research and Internationalization, Dublin City Council*

## 3.8 Designing for Sustainability

Developing a **product service system** is a core OI2 pattern to support sustainability. Product Service Systems are a new innovation pattern, which looks to move organizations from delivering products to delivering products/service and have more sustainable consumption and supply. The IOT is a fundamental enabler of this pattern, which is also sometimes called servitization. Roll-Royce's 'Power by the hour' whereby they sell hours of flight time rather than jet engines, all enabled by advanced telematics, is the most commonly referenced example of this pattern.

Equivalent to this approach is the 'extended product' approach where conceptually the product consists of both the tangible and intangible part, which often is service. This integrated approach impacts the design of the extended product, optimizing the whole value the product brings to the user.

Some auto companies are looking to see how to change their business models, Daimler's Car 2 Go car service being an example and here the motivation of the business changes from instead of maximizing sales and thereby consumption of physical resources, the goal becomes to maximize the utilization and the longevity of the assets, thereby also minimizing the consumption of non-renewable resources. In the case of Car 2 Go, the vehicles are also often electric vehicles. Broader adoption of this model requires buy-in from consumers that a car is no longer something that we need to own but that we could instead buy mobility as a service. (MaaS)

Ultimately, the collective outcomes of the OI2 efforts is that we get to a circular economy or even what Stahel calls the performance economy, whereby economic and jobs growth are achieved but decoupled from resource and environmental impact.

### 3.8.1 High Frequency, High Precision Control Systems for Societal Level Systems

Ever increasing connectivity and the ever increasing power of compute are leading to the emergence of the Internet of Things. In the future everything from cars to electrical substations to washing machines wiil be connected to the internet which will create the opportunity to introduce high frequency, high precision closed control systems to societal systems which were previously in Open Loop. For example, the electrical grid has been designed as a one-way linear system where energy is generated in bulk capacity and then distributed (quite inefficiently) through high voltage, medium voltage, and low voltage distribution systems. With the increasing availability of local renewable energy (wind, solar, etc.), smart home systems and smart heat storage systems, the opportunity exists to redesign the grid creating value for all participants, lowering costs and making the overall solution more sustainable. One Horizon 2020 project with a set of stakeholders from across the energy value chain from generators to consumers called Real Value will research and demonstrate this across 1500 homes in Germany, Ireland, and Latvia. This model is an example of the emerging concept of collaborative consumption.

At the core of these kinds of innovations are the twin ideas of systems and closed loop control through enabling functions of acquisition, analytics, and actuation. Data is acquired from a thing or system, and then analytics are performed to provide decision support, which then can drive actuation to change parameters to effect service improvements or efficiencies. The integration of these three capabilities enables the creation and operation of high frequency, high precision management control circuits. An example of such a system of systems would be a dynamic congestion charging system in a city which dynamically updates congestion charging based on parameters such as localized air pollution, weather, and traffic measurements to help optimize real-time traffic flows improve commute times while minimizing environment impact. The operation of such a system will also create a lot of big data and the use of machine learning and offline analytics can create a second-order feedback loop, which can drive further system improvements based on insights garnered.

# Chapter 4
# The Evolution of Innovation

The rapid spread of information and communications technology coupled with significant increases and lower costs of international flights and freight have created a changing innovation landscape. The EU OISPG sketches an evolution picture for the practice of innovation as shown in the figure below. In the past, much innovation came from individuals in places such as Bell Labs or IBM labs, which used a closed or vertical innovation approach (Fig. 4.1).

Henry Chesbrough identified and conceptualized the trend of Open Innovation where organizations also use external sources for ideas, technologies, and knowledge to contribute to future success. Chesbrough describes Open Innovation as consisting of five core components including networking, collaboration, corporate entrepreneurship, proactive intellectual property management, and finally a belief that R&D is crucial to the future of a company. The core philosophy underlying Chesbrough's paradigm for open innovation networking and collaboration is that innovation can be made quicker, easier, and more effective by the exchange of ideas. Chesbrough primarily saw open innovation as a way for individual companies to improve the commercialization of ideas for the benefit of the organizations involved. Connect and Develop, the Open Innovation strategy pursued by Procter & Gamble, initiated by the then CEO A. G. Laffley, is one of the most well-known examples of open innovation with Procter & Gamble reporting that nearly 50% of their new product ideas come from outside the company. The figure below shows how external ideas and technologies can transfer into a company's innovation pipeline to create value (Fig. 4.2).

Chesbrough correctly identified that Open Innovation is a bi-directional process and that unused or dormant ideas, technologies, patents, etc. can also flow out from one organization to other for exploitation.

What is however very important to notice in the model if Open Innovation by Chesbrough is that it still is based on linear innovation models, and is very much based on cross-licensing across organizations. The societal capital, creative commons, and intellectual *structural* capital of the ecosystem is very weakly taken into account in this model.

© Springer International Publishing Switzerland 2018
M. Curley, B. Salmelin, *Open Innovation 2.0*, Innovation, Technology, and Knowledge Management, DOI 10.1007/978-3-319-62878-3_4

**Fig. 4.1** The Evolution of
Innovation (source B
Salmelin, DG CONNECT)

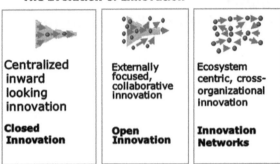

**The Evolution of Innovation**

Centralized
inward
looking
innovation

**Closed
Innovation**

Externally
focused,
collaborative
innovation

**Open
Innovation**

Ecosystem
centric, cross-
organizational
innovation

**Innovation
Networks**

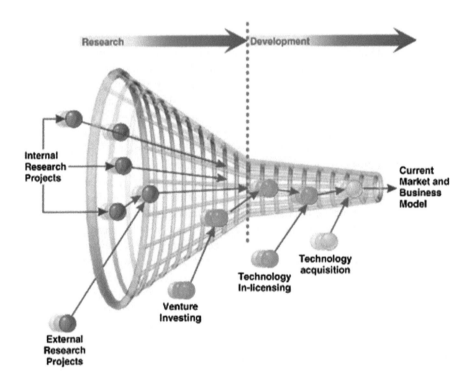

**Fig. 4.2** Open Innovation: source Baldwin and Curley, adapted from Chesbrough

Open Innovation 2.0 (OI 2.0) as defined by the EU Open Innovation and Strategy
Policy group (OISPG 2010–2016) sees the benefits of collaboration and networking
from a broader perspective as a way for firms and other organizations to improve
their innovation base so as to make optimal use of the societal capital and 'creative
commons' at their disposal. Vallat (2010) eloquently states '*In addition to exchang-
ing technology, by informal or even formal means, as in Chesbrough's ideal, the
focus here is on the involvement of all actors in the innovation ecosystem, including
end-users and end-user communities, brought together to share experience, information and best practices, and build strategic alliances and cross-disciplinary*

*collaboration'*. This type of collaboration and networking leverages the benefits of Open Innovation to a fuller extent and creates a pool of capabilities, experience, and knowledge to create the so-called creative commons. According to Vallat (2010), it favours the development of Silicon Valley Dynamics and *'Positive spill-over effects stimulated by the open environment enhance value creation for the benefit of society as a whole, and not only for the firms involved'*.

With a more open business model, there is a higher access to the market, lower costs for innovative ideas, and risk sharing. In addition, technology has a crucial role in facilitating information sharing because now it is easier to get and provide information, benefit from information sharing, and using others ideas for the greater benefit. However, there are some issues. In large companies, open innovation relates to buying or selling technologies, but for smaller companies it means collaborating and sharing technology and ideas with other parties as part of their business model. This can create an ambiguous problem and raises the question of 'how much information can and should be shared to succeed preserving the most sensitive information'.

## 4.1 Defining Innovation

While there are many accepted definitions of innovation, we will use the following definition to provide a common reference point for this book.

> *Innovation is the creation and <u>adoption</u> of something <u>new</u>, which creates <u>value</u> for the organization that adopts it.* (Baldwin and Curley 2007).

The three key words are newness (or novelty), adoption, and value. An idea of course does not have to be completely new, just new to a user, organization, focus area or, indeed, society. Indeed many successful innovations are adaptions of existing ideas, products, or services, which are already adopted and successful in some other sphere. Innovation is also about value creation and unless value is sustained, innovations will fall away. Holt (2002) says innovation is the fusion of a user need and a technological opportunity. Ultimately, there is only successful innovation when a user, organization, and society perceives and receives value.

According to Schumpeter, innovation is accomplished by 'entrepreneurs' who develop new combinations of existing resources. Intelligent recombinations of existing and emerging solutions means that innovation is often less about invention and rather more about the reapplication and synthesis of existing and emerging knowledge and solutions.

Innovation is of key concerns to politicians—UK Chancellor George Osborne showed a deep understanding of innovation in a speech to the Royal Society when he said, *'Innovation is not a sausage machine, You don't get it by a plan imposed by government and you can't measure it just by counting patents or even just spend on R&D. It is all about creative interactions between science and business. You get innovation when great universities, leading-edge science, world-class companies and entrepreneurial start-ups come together. Where they cluster together, you get some of the most exciting places on the planet. That is where you find the creative ferment which drives a modern dynamic economy'.*

As stated before key for our understanding is not the definitions but the fact that innovation is about making things happen in new ways. Often innovation definitions are strongly related to the context, whether science based, linear, triple helix, etc. We want to look at innovation from the impact perspective and see which are the important factors why Open Innovation 2.0 can make a difference in the highly dynamic innovation landscape we are in.

## 4.2   Creative Disruption

Innovation together with bank credit, according to Schumpeter, are the economic mechanisms 'that define a large part of the history of mankind' [12]. In his *Theory of Economic Development,* he classified innovation into five categories: new products (or goods), new methods of production (or processes), new sources of supply (or half-manufactured goods), the exploitation of new markets, and new ways to organize business. In Schumpeter's original schema, innovation is accomplished by 'entrepreneurs' who developed new combinations of existing resources [14]. However, in his later works, he came to regard the large corporation as the innovative engine driving the development of leading economies [15]. His emphasis of the *entrepreneur* being a single individual changed to viewing the concept as capable of being embodied by a collaborating team of people.

### 4.2.1   *Innovation and Growth*

Prof. Clayton Christensen argues that there are essentially three kinds of innovations, which have different effects on growth. They are

- Market creating innovations
- Sustaining innovations
- Efficiency innovations

Market creating innovations are the type of innovations, which create growth and jobs. Efficiency type innovations are those, which improve margins but actually destroy jobs. Christensen argues that economies like Japan have overly focused on efficiency type innovation and that this has led to the Japanese economy stagnating. One important aspect of modern innovation is to create new—products, services, and foremost markets. Due to the latter, the co-creation of new markets with the end-users as seamless process is essential. End-user involvement is necessary when exploring new boundaries irrespective of whether they come from technology, technology application, or behavioural change in the real-world settings. Often the user communities are taken fully on board at very late stage of the innovation process—too late to get valuable information on scalability and success factors of the development work.

## 4.3   Ten Types of Innovations: Full Spectrum Innovation

Increasingly, there is recognition that the linear model of science to innovation is breaking down. Doblin's ten types of innovations provide a way of thinking about innovation more holistically and how value is captured. This builds on seminal work originated by Schumpeter In his *Theory of Economic Development* book where he classified innovation into five categories: new products (or goods), new methods of production (or processes), new sources of supply (or half-manufactured goods), the exploitation of new markets, and new ways to organize business. Doblin has extended this thinking to develop a new innovation taxonomy. The taxonomy includes business model, networking/ecosystem, enabling process, core process, product performance, product system, service, channel, brand, and customer.

In the model, there are four categories of innovation levers—finance, process, offering, and delivery/experience. In finance, the business model innovation is about how you design, create, and capture value through selling product or service. Networks and alliance are about how you join forces other companies or organizations for mutual benefit. The process sector coupled with the finance element defines the configuration that underpins a sustainable innovation. The offering category refers to the product performance, product system, and service elements. Finally, the deliver/experience category focuses on the channel, brand, and indeed the experience experienced by the customer/users.

A key aspect of the Doblin's ten types of innovation is to encourage innovators to look beyond just product or performance innovation and to suggest that other areas such as business model innovation or user experience innovation may bring higher returns.

A key insight gleaned from reviewing the 10 types of innovation is that the linear model of innovation, from science to business, is too simplistic a model to use and can be ineffective and inefficient if followed to the letter of the law. Again, referring to the scope of open innovation we can argue how strongly we should focus on discovering the new with the OI2 approach, versus improving. Using an 'improving' perspective does not take advantage of the multidisciplinary nor disruptive approach where one can experiment and prototype and focus on different types of innovation to create value.

## 4.4   The Extended Innovation Value Chain

Birkenshaw and Hanson's description of the Innovation value chain as having three core components; idea generation, idea implementation and idea diffusion is useful, although it needs to be extended as shown in figure 12. The innovation value chain starts with idea generation, followed by prioritizing and funding ideas to convert those ideas to products and finally to diffusing those products and business practices. However, diffusion is not enough and a fourth value chain step needs to be added to drive adoption. Adoption, which is a function of six Us (including utility,

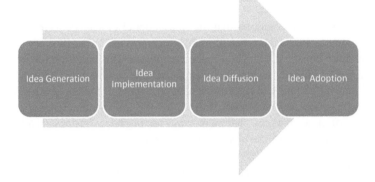

**Fig. 4.3**   Innovation Value Chain, adapted from Hansen and Birkenshaw

uniqueness, usability, usability, user experience, and design for use), is about the actual acceptance and use of a new product or service (Fig. 4.3).

Like a venture capital pipeline, the more ideas and knowledge sources that can be drawn upon the better the probability of successful innovation. '*By accessing a greater number of knowledge sources, the firm improves the probability of obtaining knowledge that will lead to a valuable outcome*' (Leiponen and Helfat 2011: 225). The so-called variance hypothesis (Dahlander et al. 2014), i.e. predicts that exposure to diverse sources of information provides the 'requisite variety' of ideas and knowledge needed to create innovations. There are important lessons from Geoffrey Moore's 'Crossing the Chasm' book about how to deliver 'the whole product' so that everything is in place to enable adoption of a new product.

Although the linear model of innovation is breaking down this is still a very useful conceptual model to think about where innovation supporting investment and policy should be focused. The inaugural OI2 conference paper (Curley and Salmelin 2013) points out that in Europe the level of funding in each phase is directly the opposite to the level of the returns, arguably this can be called the European Innovation paradox. The Aho report on an innovative Europe pointed out that average EU innovative procurement rates at 3% were very significantly lower than the government innovative procurement rates in the USA where the small business innovation research (SBIR) program helps small businesses in the US conduct research and development and empowers this sector by providing about $2.5 billion in funding per year. The recipient companies and projects need to have potential for commercialization and be aligned with the US Government R&D agenda and needs.

Although EU policy makers are finally getting to grips with implementing mechanisms for this, it points out the time lag between the speed at which the practices and methods of innovation are changing compared to the innovation policy.

Focusing on this European research investment dilemma, the characteristics of the new Horizon 2020 framework is moved from research and development towards innovation. The transition is affecting the overall design of the programme, the design of projects, creation of new support instruments as well as new criteria for successful projects. To measure the impact Innovation Radar on project level is been created as well which is a hugely important development. This allows to look at all Horizon 2020 programs from a portfolio perspective and to measure innovation success. Andy Grove, one of Intel's founders frequently said 'If you can't measure it, you can't manage it' and being able to measure innovation performance is crucial to maximizing the productivity of innovation efforts (Fig. 4.4).

The figure below shows the typical anatomy of a Horizon 2020 program where multiple organizations partner together to do joint research and co-innovate together to ultimately co-create new products and services.

When looking at breakthrough innovations, we increasingly see how linear innovation is replaced by a mash-up iterative innovation model where the innovation environment creates fruitful conditions for positive collisions across different disciplines, maturity of technologies, and the market creation. A lot of the innovation process is also experimental and prototyping by character to be able to dynamically and more quickly respond to the real needs.

With advances in global information and communications technologies, the process and practice of innovation is evolving and morphing at a very rapid pace. Innovation is a multifaceted and socially complex process where myriads of knowledgeable actors (organizations, individuals, institutions) interact and shape knowledge and resources trajectories to create value (Assimakopoulous et al. 2012). Despite David Teece's statement that '*No study of innovation can ever claim to have the last word on the subject. The phenomenon is too complex, dynamic, and adaptive to fit into a single conception for any extended period of time*', the remainder of the book attempts to create a snapshot of some of the key patterns we see emerging.

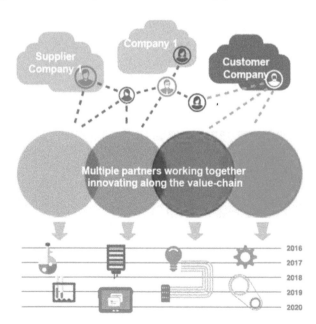

**Fig. 4.4**  Horizon 2020 project, source Spinverse

# Chapter 5
# Framing OI2

In a seminal blog in 2010, John Hagel and John Seely Brown identified that the next key challenge for open innovation was open innovation itself.[1] Hagel and Seely Brown say 'We're moving from a world where value is created and captured in transactions to one where value resides in large networks of long-term relationships that provide the rails for much richer "knowledge flows"'. Hagel and Seely Brown argue that the opportunity is to build long-term trust-based relationships amongst ecosystem players and encourage participants to build cumulatively upon the contributions of others. Karl Erik Sveiby says that trust is the bandwidth of communications. Hagel and Brown advocate an extended open innovation approach, which promotes fostering and building upon relationships between the players in an ecosystem. A core proposition is that increased interaction leads to increased knowledge, which leads to increased value.

An important factor is also to recognize that increasingly we are moving to more 'information intensive' products and services where the development cost is the dominating one, and the actual replication/multiplication and distributing cost is close to zero. Apps and 3D printing are both good examples of this. This leads to different cost and revenue model, especially when we can see that in many cases the first one in the market 'takes it all'. The long tail of for example the app economy is very thin.

This will likely also lead to multiple revenue streams of different types of interlinked products and services rather than to the old mass production dominance. Of course, there are and will be exceptions in sectors, which require extremely high initial investment in technology and production facilities. However, for example, the 3D printing technology might possibly disrupt some manufacturing companies.

In an attempt to provide a description and taxonomy of Open Innovation 2.0, key patterns of the emerging Open Innovation 2.0 paradigm are described. The goal is to help innovation practitioners and academics achieve results from innovation which are more '**predictable, probable, and profitable**' and to drive innovation impacts far beyond the scope of what any one organization could achieve on its

---

[1] http://blogs.hbr.org/bigshift/2010/02/open-innovations-next-challeng.html.

© Springer International Publishing Switzerland 2018
M. Curley, B. Salmelin, *Open Innovation 2.0*, Innovation, Technology,
and Knowledge Management, DOI 10.1007/978-3-319-62878-3_5

own. OI2 is creating faster dynamics to create new markets and to bet right based on common shareholder vision.

When we speak of 'Open', we mean openness in its broadest sense, from opening up the input to the innovation funnel to extending the innovation focus from not just product innovation but across a panorama including business model innovation, user experience innovation, and service innovation all the way up to openness of individuals, organizations, and society to embracing and adopting innovation. To this end, a key distinguishing attribute of Open Innovation 2.0 is focused on adoption and co-creation of the market based on common vision across all stakeholders.

Michael Schrage of MIT has written, 'Innovation is not innovators innovating, it is customers adopting'. While much of the focus of innovation studies is on innovation creation, very often the hardest part is the adoption of innovations. Machiavelli famously wrote in the Prince 'It ought to be remembered that there is nothing more difficult to take in hand, more perilous to conduct, or more uncertain in its success, than to take the lead in the introduction of a new order of things. Because the innovator has for enemies all those who have done well under the old conditions and lukewarm defenders in those who may do well under the new. This coolness arises partly from fear of the opponents, who have the laws on their side, and partly from the incredulity of men, who do not readily believe in new things until they have had a long experience of them'. Thus, a key aspect of adoption is developing and sustaining a culture where innovation is the norm, rather than the exception.

Edward de Bono said, 'Real innovation implies a readiness to explore and implement new ideas. But many organizations have a deep-seated fear of failure and do not like to try new things, even when much lip-service is paid to innovation'.[2]

Taking all this into account, a strong focus is on the interaction with the users who in the OI2 paradigm are genuine co-creators, and the solution providers. The more efficient this interaction is the better dynamics the market creation has, and we genuinely are moving to win-win game. If we do not focus on the creation of new markets, we have a great risk to end up with the old win-lose game against existing products and solutions. Breakthrough innovations and disruptive innovation is what we need to look for in the new landscape.

## 5.1  Design Patterns

A key aspect of Innovation 2.0 is a focus on adoption and a key new concept in helping with accelerated adoption is the concept of design patterns. Design patterns are generally reusable solutions for commonly reoccurring problems. In the spirit of walking our own talk, we use the concept of design patterns to help articulate the core mechanisms of OI2.

A design pattern is a construct, which exists in software engineering (Gamma et al. 1994), and the concept when ported to the general field of innovation can

---

[2] http://www.management-issues.com/edward-debono.asp.

prove very useful. A design pattern is a general reusable solution to a commonly occurring problem (Vaishnavi and Kuechler 2004/2005), and it is often manifested as a description or template for how to solve a problem that can be used in many different situations. Vaishnavi and Kuechler (2004/2005) specifically define patterns as a solution to a problem in a recurring context and as a general technique for approaching a class of problems that are abstractly similar. Appleton (2000; P3) defines a pattern as 'a named nugget of instructive information that captures the essential structure and insight of a successful family of proven solutions to a recurring problem that arises within a certain context and system of forces'.

In addition, Vaishnavi and Kuechler (2004/2005) describe patterns as similar to, but shorter and more structured than the case studies used in business school classes, which communicate similarly complex and subtle information. A more concise definition of a pattern is provided by Appleton (2000) as 'a pattern is a named nugget of insight that conveys the essence of a proven solution to a recurring problem within a certain context amidst competing concerns'.

Leveraging how design patterns are used in software engineering (Appleton 2000) the goal of patterns in innovation should help accelerate solutions particularly to complex problems. For complex problems or opportunities, establishing a common vocabulary or language can be difficult, and design patterns create a shared language and vocabulary for communicating insight and experience about recurring problems and their solutions (Appleton 2000). Codifying the linked solutions and capturing their relationships enables the capture of a useful and re-applicable body of knowledge (Appleton 2000).

Patterns were introduced as a design concept by Alexander et al. (1977) and the Gamma et al. book on design patterns as applied to object-oriented programming began to popularize the use of patterns in computer science. Another application of design patterns in the field of information systems was the work of Fowler et al. (2002) in the domain of enterprise application architecture.

## 5.2  OI2 Core Patterns

In the remainder of the book, we will start by discussing the core pattern of shared purpose. Then, we will discuss six core active patterns in turn including platform, ecosystem, planning for adoption, agile production, industrializing innovation, and finally data-driven innovation. We will also discuss the collective benefits and the outputs of OI2.

A pizza pattern is the metaphor we used for communicating our design patterns. A pizza is a universally recognized pattern for constructing tasty food from multiple ingredients quickly. The base of the pizza provides the platform for the chef to assemble and arrange multiple ingredients so that a 'solution' can be generated to meet different criteria such as nourishment, taste, and flexibility. A pizza can be consumed by slice or multiple slices of pizza from different types of pizzas; it can be consumed in any one meal. Similarly, we present the initial OI2 patterns as pizza

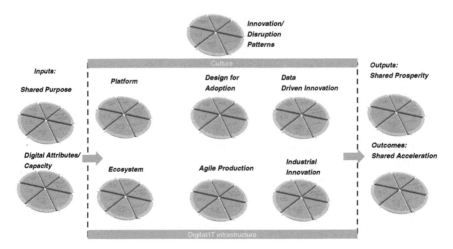

**Fig. 5.1** OI2 Core Patterns

patterns with the concept that they can be assembled and combined in a way that makes sense for the ultimate consumer of the solution. In the spirit of OI2, Peter Finnegan, South Dublin Area Manager suggested the metaphor of a pizza as a good way of communicating the concept of the OI2 design patterns.

The patterns are arranged as depicted in the (Fig. 5.1) with shared purpose and digital technologies being the core input patterns and outcome and output patterns residing outside of the OI2 innovation factory. In the Innovation black box we describe six patterns—platform, ecosystem, designing for adoption, agile production, industrial innovation, and data-driven innovation. Different kinds of innovation lenses can be applied to this innovation factory, and in this framing we use patterns of digital disruption as a lens.

A core pattern is that of a **platform**. This could be just an agreed shared taxonomy but more often than not will be a technical platform, which facilitates co- and derivative innovation. A platform is a set of core components from which a pipeline of derivative products or services can be efficiently created and launched. A pivotal characteristic of the most successful platforms is that they enable and experience networks effect, whereby the more users and/or complementors the more valuable the platform becomes and the greater value it creates. Thus, platforms when they reach a certain critical mass, exhibit a positive feedback loop where users get more benefits as more users join or as more organizations or individuals develop complementary products or services. Uber and Airbnb are examples of platforms, which use collaborative consumption as a core principle growing their business but also enabling better sustainability of both the production and consumption parts of the value chain.

Associated with a platform is an **ecosystem**, which drives the inertia around the platform, co-developing innovations, and distributing and driving adoption of the outputs. Leveraging the platform pattern is the concept of the reverse innovation pyramid where instead of innovation something done to users, users are part of the innovation process, and indeed sometimes take the lead. The Apple iPhone and app store is a good example of an ecosystem and associated platform where hundreds of thousands of developers have helped create and sustain a winning ecosystem and set of products. Another good example of an ecosystem/platform coupled with the reverse innovation pyramid is the Lego Ideas platform. Lego encourages children and others to submit new product ideas for consideration by Lego. The process is simple but not overly easy with participants submitting ideas, then required to get enough support, 10,000 supporters for their product ideas before Lego reviews it and ultimately decides to productivize the idea or not. There is shared value created with successful inventors getting 1% of product royalties. Another example is Organicity a Horizon 2020 smart cities project in London, Santander and Aarhus where 25% of the total budget is allocated to citizen-initiated projects.

Most innovation value comes from adoption (80%) rather than creation efforts (20%) (OECD); however somewhat bizarrely most innovation efforts are focused on creation activities. A core-distinguishing feature of OI2 is a focus on adoption, involving co-innovators and users so that innovations are ***designed for adoption*** by focusing on utility, ease of use, user experience, and the creation of network effects. The use of **agile development and production** methods is another core pattern of OI2 so that a non-linear iterative, higher velocity development process replaces the linear waterfall development model. **Industrial Innovation** itself is a core pattern so that it becomes a repeatable, scalable, and automated system—this means thinking about innovation as a capability to be managed and invested in. Finally, a 2017 edition of the economist argued that 'Data is more valuable than oil'. The data-driven innovation chapter describes common patterns for deriving value from data, which is also a core OI2 pattern.

In an attempt to provide a snapshot description and taxonomy of Open Innovation 2.0, key patterns of the emerging Open Innovation 2.0 paradigm are described in the remaining chapters. The goal is to help innovation practitioners and academics achieve results from innovation, which are more 'predictable, probable, and profit-able', and to drive innovation impacts far beyond the scope of what any one organization could achieve on its own. OI2 is creating faster dynamics to create new markets and to bet right based on common shareholder vision.

# Chapter 6
# Shared Purpose

Open Innovation 2.O (OI2) is a new paradigm based on principles of integrated collaboration, co-created shared value, cultivated innovation ecosystems, unleashed exponential technologies, and rapid adoption, often accelerated by an innovation methods based on prototyping and experimentation in real world.

At the core of OI2 is the concept of shared vision/value and the quadruple helix innovation model where government, industry, academia, and citizens/users aligned around a common shared vision, can drive structural improvements far greater than any one organization could achieve on their own through collaborative innovation. Cultivating and orchestrating an ecosystem, which leverages a common innovation platform allowing co-creation and deep user involvement and innovation, is crucial to successful results. The figure below shows a depiction of the Vision of a Digital Europe that help us in Intel Labs Europe (ILE) drive good alignment with many partners (EU Commission, Governments, Academics and Industrial researchers) around an innovation agenda where digital technologies help transform different parts of European society such as cities, healthcare, and transportation. The cultivation of such an ecosystem based on a shared vision allows the alignment, amplification of resources, and acceleration of results (Fig. 6.1).

## 6.1 Shared Purpose

Central to the concept of Open Innovation 2.0 is the idea of shared purpose when shared vision, shared values, and value as espoused by Peter Senge, Porter and Kramer (2011), Curley (2004), and others are critical. This foundational pattern focuses on creating a shared vision to align the energy, investment, and interests of all parties participating in the ecosystem or transformation initiative. Additionally, this pattern helps to encourage organizations to explicitly document the shared

© Springer International Publishing Switzerland 2018
M. Curley, B. Salmelin, *Open Innovation 2.0*, Innovation, Technology, and Knowledge Management, DOI 10.1007/978-3-319-62878-3_6

**Fig. 6.1** ILE Digital Europe Vision, source Curley/Fleming

value they hope to create so that win-win scenarios are created. When an ecosystem's members have shared values, difficulties and problems, which are inevitably encountered, are often more quickly and satisfactorily resolved or indeed totally avoided. Typically, transformation efforts require significant energy and time – investments and efforts from different organizations need to be aligned as best as possible. Different organizations will have different motivations, goals, success criteria, and dissonance that will naturally occur unless interests are aligned around a shared vision.

When interests, investments, and energies are aligned, the result should be alignment, amplification, and acceleration of results as well as attenuation of risk. Rather than participants complying to a schedule or program of work they are committed to a vision and will typically deliver more than would otherwise be the case. Brainstorming and collaboration to create a compelling vision is often the first step. Recognizing the key problem that needs to be solved or key research question to be answered provides an essential input to creating the vision. Crowdsourcing the vision statement to the ecosystem to produce a number of candidates for consideration as an agreed shared vision is often productive. For example, when Boris Johnson commissioned the Smart London Board, the exercise of creating a vision was outsourced to ordinary Londoners with several hundred citizens of London submitting proposals. Ultimately, one such submission was used as the rallying vision for the Smart London efforts—'Using the creative power of technology to improve the lives of ordinary Londoners'.

Having a shared vision, value, and values will maximize the chances of success of a consortium or ecosystem in delivering a solution for a significant challenge and will help deliver that solution in the most efficient and effective way (Fig. 6.2).

There are six core elements to the OI2 comprehension/pattern of shared purpose.

**Fig. 6.2**  Shared Purpose

- Shared vision
- Shared value
- Shared values
- Shared value at risk
- Shareholder and stakeholder value
- Stored value

### 6.1.1   Shared Vision

Victor Hugo wrote there is nothing as powerful as an idea whose time has come. Major structural changes and improvements can be achieved when there is a powerful and compelling shared vision. John F Kennedy's vision of putting a man on the moon and bringing him home safely is a great example of a vision, which helped mobilize the USA to land a man on the moon while also creating a significant number of innovations, which could be harvested and used elsewhere in society. An example of the commitment, which such visions evoke, is the famous response of the NASA janitor, when asked what his job was, he replied 'I'm working to put a man on the moon'. A shared vision is one to which both individuals and organizations commit to because it reflects their own personal and organizational visions, respectively.

### 6.1.2   Shared Value

Shared value is about reconceiving the intersection between society and corporate performance, seeking win-win outcomes and being profitable through solving big problems. Underpinning the idea of shared value is a shared vision of what can be achieved. Contributors to Wikipedia have signed up to Jimmy Wales's vision of a new kind of encyclopaedia and contributors get shared value from using the encyclopaedia. A key tactic is producing digital shared value. According to Carlos Härtel, head of GE Research Europe 'A 1% digitally enabled fuel saving in the conventional aviation industry is worth $30 billion a year, and the company that enables could have half of that game'.

Intel Corporation's vision under the leadership of Paul Otellini's is an example of Porter's shared value—'This decade Intel will create and extend computing technology to connect and enrich the lives of everyone on the planet'. Intel then believed that it had both a responsibility and opportunity to connect and enrich the lives of everyone on the planet. As is wont to happen when a new CEO came on board, this vision replaced by one which was different and less shared value oriented—'If it computes, it runs best of all on Intel'.

Porter and Kramer's idea is also an extension of C. K. Prahalad's thinking when he wrote about 'the fortune at the bottom of the pyramid'. Prahalad articulated a theory that companies can make significant profits by solving problems for people living in the so-called Third World. Thus, the possibility exists to do noble work as well as being profitable. Ultimately, this kind of thinking will not only improve people's lives and drive profits but also in parallel expand collective human potential. Ajay Banga, CEO of MasterCard has a mantra of 'Do Well and Do Good' with an objective of driving financial inclusion, improving the quality of life of people who had no access to financial services and growing the overall industry.

Albert Einstein summarized the necessity of shared value thinking when he said, 'It is not enough that you should understand about applied science in order that your work may increase man's blessings. Concern for man himself and his fate must always form the chief interest of all technical endeavours'. This links to the core idea of OI2 that organizations and people innovating and transacting in an ecosystem should have shared values.

### 6.1.3  Shared Values

Values are important and durable ideals or beliefs held by members of a culture, company, or community about what is good or bad and desirable or undesirable. They heavily influence both an individual's and collective behaviour, posture, and attitude and serve as general guidelines for dealing with various situations. For example, companies like Mastercard, General Electric, and IBM who have endured large changes typically have enduring values, for example, focused on risk taking and customer orientation, which help guide behaviour and responses to different challenges and decision-making processes. In a particular ecosystem, it is extremely helpful if all participants hold similar shared values, and this can lead to the output of the ecosystem being strongly synergistic rather than members of the ecosystem or community working against each other in certain situations. It is also important to see new work structures in this context. The competencies of individuals will increasingly be shared across multiple actors/firms based on the values the worker and the organization have. The organizations who best capture the empowered customers, citizens, and the high-value competencies will more likely be the winner in the end of the game.

**Case Study: Innovation Value Institute**

The Innovation Value Institute (IVI) based at Maynooth University, which celebrated its tenth anniversary in 2016 is an excellent exemplar of OI2 in action and how OI2 supported by vision and resilience can deliver multiplicative returns. From very humble beginnings, IVI is now taking its place amongst the leading Information Systems Research organizations in the world based on the adoption of its research by organizations. The Innovation Value Institute was founded in 2006 by Professor Martin Curley, then of Intel and Professor John Hughes, then President of Maynooth University with a vision of creating research outputs, which could drive a structural change in the way organizations get value from information technology. Supported by many world leading companies such as BCG, EY, Chevron, BP, SAP, and Microsoft the institute quickly developed momentum using an OI2 and engaged scholarship approach. Initially, the institute was funded primarily from the participating companies and leading global companies, even competitors of each other, who bought into the shared vision of working together to create a framework and set of practices, which could drive a structural change in the way organizations get value from IT.

In 2009 under the Irish Technology Centres Programme, Enterprise Ireland announced a 5 year/€5M Phase 1 investment in the Innovation Value Institute (IVI) Technology Centre. The main aim of the IVI was to create a set of tools and processes housed inside an assessment framework that would give IT organizations the means to substantially upgrade their IT capabilities across the board, making them leaner, more efficient, cost-effective, and better able to serve their client base both internally and externally.

The IVI collaboratively developed and released the IT-Capability Maturity Framework, which is becoming a candidate for the gold standard for the management of IT Value and IT-Enabled innovation. This framework has been embraced by companies globally and from many sectors including oil & gas, banking, construction, big pharma, transport, and IT with over 500 organizations having used the IT-CMF.

By any measure, the performance of the Innovation Value Institute has been strong with over a quarter of million hours of research in an engaged scholarship mode being contributed by industry executives and specialists and the IVI research team with over 1400 professionals contributing to the knowledge base and over 300 publications and whitepapers. Already 800 professionals have been certified on the IT-CMF and in October 2015 the IT-CMF Body of Knowledge Guide, a 700+ page tome was published by IVI as a reference for IT executives and professionals worldwide. Key to the ongoing success of IVI was creating a shared vision and platform to enable co-creation and development and governance/orchestrating an ecosystem to co-develop and help diffuse and drive adoption of the IT-CMF. The ability to create shared value and the common values of the different ecosystem participants delivered very strong results.

### 6.1.4  Stored Value

Stored value is an increasingly important concept in that there is latent or stored data created by the interaction of community members both in the creative and execution phase. Choosing vibrant members for the ecosystem who have high potential will help subsequent value delivery. An example of the recognition of stored value is that the Smart London Plan identified that Londoners generate the data that helps London manage its transport, social, economic, and environmental systems. Digital technology presented opportunities for London to use this data to function better, and for Londoners to help shape and be a part of the solutions.

### 6.1.5  Shareholder Value

Commercial organizations are ultimately gauged by ongoing contributions to increased shareholder value. Digital innovation efforts must be tied to the probability of increasing shareholder values. Tools such Osterwalder's business model canvass and the value dials practice developed at Intel are useful in trying to tie-specific innovation investments to unit improvements in core variables, which have a material impact on the business. Organizations such as the Depository Trust Clearing Corporation (DTCC) which itself plays a fulcrum role in using value dials to help select and then manage IT and digital innovations to help improve the probability that IT and digital innovation investments ultimately deliver on the promise that was initially proposed.

#### 6.1.5.1  Value Dials

Intel IT originally developed value dials, a framework for measuring the business value contribution of IT projects. Value dials are financial metrics that map business value to an organization's top and bottom line. For each investment project, a number of value dials were identified, targeted, and measured for improvement as the solution was developed and deployed. Each value dial was specified using a specific metric or equation that specified the impact. Value dials are typically the business variables, which make a difference to the profitability of the business. An example of a value dial would be 'days of inventory'. Reduction in days of inventory reduces the cost of carrying inventory and can be achieved by implementing just-in-time systems, faster manufacturing throughput system, or better market intelligence, for example. Another example of a value dial is 'days of receivables' and solutions which enable a company to receive payment faster improve cash flow and allow better cash flow management. For a bank, investments, which target improvements in cost/income ratio, are particularly sought after.

Once a comprehensive set of value dials is available and maintained for an organization, these can be used to allow innovators define, design, and deliver for business value. The use of value dials can be used to allow comparison between different innovations competing for investments. I developed a tool at Intel called the Business Value Index which looked at early stage innovations as 'options' and assessed the likely potential value of innovations based on three vectors of contribution business value, IT efficiency, and financial contribution.

## 6.1.6  Shared Value at Risk

Value at Risk (VaR) is an important measure of the risk of investments but also the measure of exposure if a specific regulation is not complied with. For example, a bank, which does not comply with the European Commission's General Purpose Data Regulation, risks significant fines, which are materials to the operation of their business if they are found to have breached data collection regulations. The concept of VaR can be extended to an ecosystem and its constituent organizations so that there is a strong value at risk if their ecosystem or collective investments are not successful or surpassed by those of another ecosystem which reaches critical mass first and establishes a dominant design. When a network effect occurs, it is often very difficult to displace an incumbent from their position, which gets increasingly stronger as more users and partners adopt the services. The concept VaR is often used by regulators and financial organizations to assess the amount of assets which might be required or held to cover possible losses and additionally can help ecosystems assess the amount of investments and options they need to hold to help avoid catastrophic losses due to being outpaced and out-innovated by a competing ecosystem.

**CERN Case Study**
CERN, the European Organization for Nuclear Research, was founded in 1954 and is an early example of OI2 at work with over 20 countries contributing to a shared research agenda to research and probe the fundamental structure of the universe. While the core research agenda for CERN is Nuclear Physics, it is perhaps more famous for a spill over innovation which came out of CERN. Tim Berners-Lee, a scientist at CERM, invented the World Wide Web (WWW) in 1989 and it emerged out of a need to enable information sharing across universities and research institutes around the world. The Web's development was enabled by the 'Open' approach, which CERN took, putting the software in the public domain on April 30, 1993, and the subsequent next release via open licence. This became both the enabler and catalyst for changing the way the world shares information and provided an infrastructure, which would underpin a wave of digital entrepreneurship. The simplicity of the WWW idea as well as the royalty free availability enabled its rapid adoption and further development.
  *CERN Open Lab as an OI2 exemplar*
  The CERN Open Lab is a microcosm of CERN and is another great example of OI2 in action focused on shared purpose. Established in 2001, the CERN Open Lab is a unique public–private partnership that accelerates the

development of cutting-edge solutions for the worldwide LHC community and wider scientific research. The unofficial mantra of the CERN Open Lab is 'you make it, we break it' meaning the Open Lab community pushes new emerging technologies to the extremes, for the extreme operating demands of the CERN Large Hadron Collider and in doing so also discovers the operating boundaries of new high tech products and services coming from the ICT industry.

In the CERN Open Lab, CERN collaborates with leading ICT companies and research institutes, provides access to its complex IT infrastructure and its engineering experience, and occasionally extends this access to collaborating institutes worldwide. Testing in CERN's demanding environment allows CERN to assess the merits of new technologies in their early stages of development for possible future use, provides the ICT industry partners with valuable feedback on their products while providing academic research institutes unparalleled access to domain and industry expertise and products. This ecosystem provides a neutral ground and platform for carrying out advanced collaborative.

The longevity and growth of the CERN Open Lab, now over 15 years in existence demonstrates the shared value created by participating collaborators. The previous CERN Director General Rolf Heuer acknowledged that the Open Lab delivered a long list of technical achievements, which played a vital role in the discovery of the Higgs Boson particle. As part of the shared purpose, the CERN Open Lab also demonstrates a significant commitment to education, with funding provided for young researchers to build the pipeline for the next generation of talent. Rolf Heuer also highlighted that the work of the Open Lab demonstrated the virtuous circle between basic and applied research at work.

The CERN Open Lab practices at the frontier of digital innovation and currently focuses research on topics such as data acquisition, computing platforms, data storage architectures, compute provisioning and management, networks and communication, and data analytics.

# Chapter 7
# Platforms

A foundational enabler and design pattern of OI2 are platforms. Platforms and their associated ecosystems are hugely powerful, and their emergence is driven by continued rapid evolution of computing and communications, which enable the digitization of many things including products, services, business models, user experience, etc. The growth of an ecosystem around a platform moves the locus of innovation from the platform owner to a network of other organizations and indeed individuals and allow operation on a scale that can be massive and far beyond the scope of what any one organization could achieve on its own.

Gawer and Cusumano (2013) define external platforms as products, services, or technologies that act as a foundation upon which external innovators, organized as an innovative business ecosystem, can develop their own complementary products, technologies, or services. A platform is a set of standard and modular components or parts from which a stream of derivative innovations can be developed (Meyer and Lehnerd 1997). Reuse of components of course enables economies of scope and scale allowing organizations to take advantage of silicon, software, and network economics. However, perhaps more importantly, the components form a foundation of primitives, which allow multiples of value to be created on top of the platform. An important attribute of successful platforms is that they evolve as environmental conditions change. A platform ecosystem consists of stable core modules, processes, variable complementary processes, and modules. The unit of competition has changed from the product to the platform. The output of a platform ecosystem is a set of complementary products and services, built on the standard core components that significantly augment and expand the capabilities of the platform. There is a trend to migrate towards platforms in a variety of industries.

The smart architecture of a platform should provide the blueprint for mass coordination and mass collaboration. Platforms come in many different shapes and sizes but overall the goal is to provide an infrastructure, which enables interactions between different types of communities, often producers and consumers. One can do this by architecting reasons to collaborate or creating incentives that repeatedly connect and pull in participants to the platform. Platforms provide a central or distributed infrastructure upon

© Springer International Publishing Switzerland 2018
M. Curley, B. Salmelin, *Open Innovation 2.0*, Innovation, Technology, and Knowledge Management, DOI 10.1007/978-3-319-62878-3_7

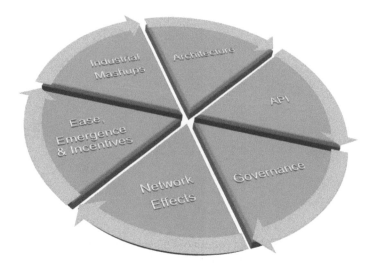

**Fig. 7.1** Platform Patterns

which participants create and exchange value. In execution mode, a platform will often match participants with each other and with content and services created or distributed through the platform. Platforms allow innovation and execution on a scale far beyond what is achievable by any one organization on its own. An important characteristic of platforms are that they are often multi-sided, where platform can bring two or more distinct groups of stakeholders together. As one example the Apple App store and iOS platform brings together end-users and app developers on either side of the platform.

A platform could be something as broad as the high tech campus in Eindhoven, which has evolved from a closed Philips Lab of about 2000 researchers to a thriving environment of more than 140 companies with over 10,000 employees. A platform could also be the Innovation Value Institute (www.ivi.ie) which provides a common taxonomy and research infrastructure for different organizations to contribute to an overarching CIO framework to drive a structural change in the way organizations create value from IT. An internal platform as defined by Gawer and Cusumano (2013) is a set of assets organized using a common structure from which a company can effectively and efficiently develop and produce a pipeline of derivative products and services. An external or industry platform is a set of assets that have been assembled to act as a foundation upon which external innovators, organized formally or informally as a business ecosystem, can develop their own complementary services or products. A key concept around platforms is the efficient use and reuse of modules or assets, which act as a foundation for derivative innovation (Fig. 7.1).

## 7.1  Architecture

The architecture of a platform is foundational to its success what is the functionality provided, what are the building blocks, and what are the interfaces provided. In a software platform, the interface is very often an Application Programming Interface (API), and this forms the basis of how other app developers can integrate, synthesize, and further develop higher level services based on the foundational assets. Facebook, Google Maps, and Amazon AWS are examples of platforms, which provide APIs for further service development. The real power of a platform is demonstrated by the Apple App store where hundreds of thousands of developers develop mobile applications based on the Apple operating system iOS and App store standards, there is a revenue share between the developers, and apple for each App sold. With exquisite hardware and software and a platform approach, Apple has created a winning platform and growing ecosystem.

Thus, a platform also requires an 'architecture of participation' to help grow the ecosystems. Particularly, in the case of a software platform, app developers must both motivate and be able to innovate on top of the platform architecture. The key instrument for enabling innovation on a platform is Application Programming Interfaces (APIs).

## 7.2 Application Programming Interface

We are heading for an API-based economy where machine-to-machine or computer-to-computer collaboration will be a dominant trend. APIs are the building blocks and scaling agenda of the API-enabled digital economy with an API offering a task or data to another computer program to use. APIs are machine readable in real-time rather than designed to be used by users, other than application developers who use APIs to create more complex functions using software. There are three kinds of APIs, private, partners, and public APIs. A private API is used within an organization to improve the efficiency and effectiveness of the software infrastructure and applications of a firm. Partner APIs enable the integration and coupling of software and business processes between an organization and a partner organization. Finally, public APIs allow anybody to use functions from an organization in an integrated way and typically, an organization uses APIs to expose functions, data, and tasks from their systems in an efficient way and in a way that can be monetized. APIs change the way that software gets delivered and make it much easier to synthesize different software components into more complex functionality.

Enabled by APIs, successful platforms exhibit non-linear growth and successful platforms treat APIs as products and evaluate their success using critical success indicators. Facebook grew by 160% against 5% growth of the once dominant MySpace 1 year after Facebook opened up its APIs to enable developers to use its platform. This statistic is very clear evidence that open API's work.

## 7.3 Governance

The governance of a platform needs to be thought through and can be explicitly codified into a software platform or may require explicit committee and a board to oversee. For a research consortium and platform like the IT-Capability Maturity Framework at the Innovation Value Institute (IVI), a consortium board made up of leading consortium members oversees the workings of the consortium while a technical committee of leading architects and researchers governs and oversees the evolution of the IT-CMF platform and also enforces and oversees quality. The more 'sides' a multi-side platform has the greater the likelihood that more formal governance is required. The IVI brought together six types of distinct stakeholders together, government, ICT industry, consultants, ICT enterprise users, analyst companies, and academia and thus required a governance process in addition to the shared vision that was created to drive the consortium. For commercial ecosystems, the governance arrangements are often codified into the developer terms and conditions, an example of which can be found at https://developer.apple.com/terms/. Such agreements determine the technical specifications for participating and contributing to the ecosystems as well as the revenue sharing arrangements.

## 7.4   Networking and the Network Effect

Networking is at the core of Open Innovation 2.0, and it is a socioeconomic process where people interact and share information to recognize, create, and indeed act upon business opportunities. Diane Bryant, Intel Senior Vice President, writes the word networking as 'netWORKING' to emphasize that although netWORKING is a social process it is also a key *work* activity used to establish and extend relationships while also enabling the creation and striking of options. Networking is very much a collaborative process. De Bresson and Amesse (1991) wrote, 'that no firm, large or small can innovate or survive without a network'. Any yet few organizations explicitly manage their network with a deliberate strategy and indeed a focused network manager.

We have all seen cases of collaboration that create effects which are at best additive, delivering a sum of the parts which is less than the sum of each of the individual components. OI 2.0 generates synergies and network effects rather than just additive effects. Synergy describes two or more entities interacting together to produce a combined effect greater than the sum of their separate effects. The root of the word synergy is in the Greek word 'synergia' which means, 'working together' and this term is one way of effectively describing one key attribute of OI 2.0. A network effect is the effect that a user of a service has on the value of that service to other people, with the telephone being the classic example of the network effect. Where there is a network effect, the value of the service is dependent on the number of people using it or contributing to it. New users, players, or transactions reinforce existing activities, and there is acceleration in the number of users and value creation. Bob Metcalfe postulated Metcalfe's law and said that the value of the network was proportional to the square of the number of connected users. David Reed, one of the founders of Lotus, postulated that the value of the network will come from what he calls 'group-forming networks', where users come together to collaborate and generate wealth. He expressed the value arithmetically as two to the power of

**Fig. 7.2** Synergy versus
Additive effects (inspired
by W.M. Gore)

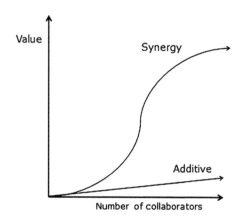

the number of users. As the number of users gets very large, the primary value will come from collaborative groups (Fig. 7.2).

In Open Innovation 2.0, the goal is not to design additive effects but to design for exponential effects where 'power' laws manifest themselves. A power law is a description of an interaction where there is a mathematical relationship between two variables, often where the frequency of an occurrence varies as a power of some attribute of another variable.

The every expanding developer ecosystem that Apple established around the iPhone is a contemporary example of a network effect and many argue that this provides the greatest stickiness to the Apple products and services. As networks proliferate and companies such as Intel and Cisco see pathways to connecting everybody on the planet, the opportunities to create network effects increase dramatically as the cost to connect to or contribute to a new service continues to fall and the access barriers drop. This is an unparalleled opportunity for would-be entrepreneurs as access to users and computing and communications costs continue to fall significantly and a global business can be built very quickly from a bedroom or garage—e.g. Facebook and Google. The European Internet foundation have identified mass collaboration as the dominant paradigm for next decade and organizations that can create platforms and venues to harness and exploit this trend can achieve great results.

A network effect occurs when a platform becomes more useful with each subsequent user that is added to the platform or system. The classic example of a network effect is the telephone system whereby as each new user is added the more valuable the network becomes. Once a system hits critical mass, a positive feedback loop drives increasing returns as more users are attracted. Often with platforms, a dominant design emerges which becomes ever more dominant due to the positive feedback loops that are associated with platforms. Network effects can be either single sided or multi-sided. A same-side network effect occurs when adding a user to one side of a platform increases the value to all participants on the same side of the platform—e.g. adding another user to Skype makes the platform more valuable to all other Skype users. Cross-side network effects occur when adding an additional participant on one side of a platform increases value to participants on the other side of the platform. For example, adding a developer to the App store side of a platform increases value to all the users of the App store. In many cases, positive network effects occur with platforms but it is also important that negative network effects can occur when, for example, too many users create congestion for other users of the system.

## 7.5  Ease, Emergence, and Incentives

Well-architected platforms become blueprints for mass collaboration (Tiwana 2014). They have to be easy to use and have to provide incentives or reasons why actors continue to return to the platform to create and exchange value. Emergence is an especially important characteristic of platforms that are successful. Emergence is a self-organizing ecosystem-wide order (Dougherty and Dunne 2011) that arises not only from preconceived design principles of the platform owner but also from the interdependent actions of platform contributors and users based on the shared vision of the platform and the continuous adaption of what participants contribute and use based on what others are doing with the platform. Where a shared vision exists and shared value is defined, the emergence of ecosystem innovation can be accelerated and more efficient compared to an ecosystem where there is no shared vision and shared value espoused.

## 7.6  Industrial Mash-ups

EY's Jeff Liu has proposed the idea of industrial mash-ups as a new way of collaborating, which will accelerate innovation in the API economy. According to EY, industrial mash-ups are new forms of alliances that will form live and die very dynamically based on synthesis of APIs to deliver customized functionality. The concept emerges from the same principles that are underpinning the sharing economy, utilizes shared data, services, or indeed physical assets increasingly via machine-to-machine methods, and enables derivative innovation to create new business value which is distinct from the original value derived from an asset, data source, or app, and allows integration of partners functionality serviced via APIs into your own organization services.

Jeff Liu (2016) explains, 'Fundamentally, the sharing economy exists because of the idea that you can separate, or detach, new kinds of value from an underlying physical thing. You're able to use your car as a transportation service, or use your apartment as a hospitality service, because another organization has built an easy-to-use, automated transactional environment on the Internet with new kinds of usage terms that depart from standard lease contracts'.

As more assets and products are made more visible and accessible via APIs, industrial mash-ups enable existing processes, assets, and IP to be monetized in new ways. Increasingly, design patterns for industrial mash-ups will become available, so that once a new form of industrial mash-up is developed; others will be able to imitate it using their own proprietary and/or ecosystem assets.

An industry where this is likely to happen quickly is in banking where the EU's Payment Services Directive 2 (PSD2) calls for opening up banks' customer information, subject to permission, by new kinds of financial players to drive innovation.

## 7.7   Multi-Sided Platforms

A core, if not the core property of a platform is that it connects two types of players together, in a way which has much lower connection and transactions costs than there would be if the different parties had to connect, find each other, and transact with each other. Typically, a multi-sided software platform brings end-users and app developers together but platforms have existed before the emergence of global software platforms as iOS and Android. For example, the credit card companies connect cardholders, merchants, and banks together. In this case and in many other cases, the credit card platform creates value by mediating interactions between different sides of a market. Of course, Global Distribution Systems such as Amadeus have long existed as platforms connecting travellers and airlines.

A winning platform exhibits network effects whereby the additions of more users create incremental value to existing users and of course to the platform. Once a platform hits a critical mass participation, growth of the platform and service can be autocatalytic as new apps or users are added to a platform, the platform becomes more value adding and attractive to more app developers and users. As with the case of autocatalytic chemical reactions, the output of a platform, which demonstrates a network effect, is fundamentally non-linear. In platforms, one can observe power laws where one quantity such as the number of users varies as a power of another, such as the number of apps.

Once a network effect is triggered, the platform and associated participants enter a self-reinforcing cycle, adding significantly more value but also creating a high barrier to entry or would be disruptors. Platforms can exhibit both same side and cross-side network effects. When one more user joins, a telephone network or a system such as Skype value is increased to all users. Cross-side network effects occur when an increase in either side of the platform creates value for both sides. For example, the more people that buy iPhones, the more that app developers are encouraged to write apps.

# Chapter 8
# Ecosystem Orchestration and Management

*'The new leaders in innovation will be those who figure out the best way to leverage a network of outsiders'.*

*Pisano and Verganti*

An ecosystem can be defined as a network of interdependent organizations or people in a specific environment with partly shared perspectives, resources, aspirations, and directions (Andersson, Formica and Curley 2010). The ecosystems with the biggest critical mass and the greatest velocity will have the most momentum and will ultimately win. The following figure shows different participants in and different formats of innovation (Fig. 8.1).

In this chapter, we will discuss some of the key patterns associated with ecosystem management and orchestration as shown in the following figure (Fig. 8.2).

## 8.1 Partnering: Triple/Quadruple Helix Innovation

A core characteristic of the OI2 paradigm is used of the quadruple helix model where government, industry, academia, and civil participants work together to co-create the future and drive structural changes far beyond the scope of what any one organization or person could do alone. When all participants commit to a significant change such as transforming a city, or an energy grid, by collaborating everyone can move faster, share risk, and pool resources. Industry, academia, and civil participants work together to co-create the future and drive structural changes far beyond the scope of what any one organization or person could do alone. When all participants commit to a significant change such as transforming a city, or an energy grid, by collaborating together everyone can move faster, share risk, and pool resources.

Triple Helix innovation involves government, academia, and industry working together to drive structural innovation improvements beyond the scope of what any one organization could achieve on its own. Henry Etzkowitz, the originator of the term, postulated that because this interaction is so complex it needed a triple helix rather than the double helix of DNA to describe it. In a generative knowledge economy, industry is seen as the locus of production (products or services), governments provide a stable and defined regulatory environment as well, often as invest-

© Springer International Publishing Switzerland 2018
M. Curley, B. Salmelin, *Open Innovation 2.0*, Innovation, Technology, and Knowledge Management, DOI 10.1007/978-3-319-62878-3_8

# Innovation Ecosystem

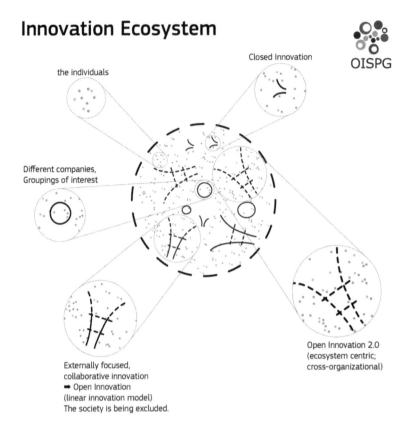

**Fig. 8.1** Innovation ecosystem constituents

**Fig. 8.2** Ecosystem

ments and investment incentives while the role of universities is changing from primarily providing a supply of trained people and education to also providing primary knowledge to the innovation process. Triple helix innovation is all about designing for and achieving systematic synergies. Quadruple helix innovation extends the concept to include the user, customer, and indeed citizens in the innovation process. The following figure depicts the collaborative nature of OI2 taken in a European context (Fig. 8.3).

Europe's FP7 and now Horizon 2020 are arguably the world's largest open innovation and research funds and Horizon 2020 is increasingly adopting an OI2 posture. In Horizon 2020, we are seeing quadruple helix innovation configurations (Carayannis and Campbell 2010; Curley and Salmelin 2013/2014) arising where the use of users and citizens as co-innovators and participants in Living Labs is actively promoted and incentivized. The following figure depicts the intertwining efforts and energies of the different partners working together towards a shared vision.

While I was at Intel Labs Europe, we invested heavily in FP7 and had more than 70 FP7 projects and then over twenty five projects were active in Horizon 2020. Intel's open labs in Ireland, Munich, and Istanbul served as portals to the broader network of more than fifty European R&D labs. Additionally, Intel had co-labs with companies such as BT and SAP where using a shared agenda we are able to move much quicker together (Fig. 8.4).

## 8.2  Citizens as Innovators

Increasingly, citizens have an appetite to participate in innovation and are an example of stored potential in an ecosystem whereby with the right platform and incentives their latent ideas and energy can be converted into kinetic energy for the good

**Fig. 8.3** Quadruple helix innovation in action in Europe

**Fig. 8.4** Quadruple helix innovation

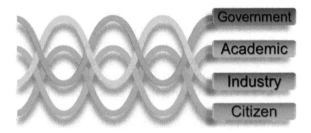

Government

Academic

Industry

Citizen

of all. Nambisan and Nambisan (2013) from the University of Wisconsin call out four different roles of citizens in the public service innovation process.

- Explorer
- Ideator
- Designer
- Diffuser

As an explorer, citizens can explicitly discover and define existing and emerging problems and opportunities in public services. As an ideator, citizens can suggest and formulate potential solutions to these and other problems in public sector. Challenge.Gov is one noteworthy example where the US government uses online contests to solicit ideas from citizens to help solve problems. As a designer, citizens can design and conceptualize solutions to specified problems with NYC Big Apps being one such vehicle for this. Finally and importantly, citizens can support the diffusion and adoption of public service innovations.

The Nambisans call out two of the core OI2 patterns to enable co-creation between citizens and government, the innovation ecosystem and platform. The innovation platform is the venue for citizen engagement and co-creation and includes ways to facilitate knowledge sharing and collaboration amongst participants and most importantly provide standard components and modularization of the problem-solving process. The ecosystem refers to the milieu and organizing structure for government agencies and citizens through creating a shared vision and worldview and defining the participative architecture to coordinate innovation and collaborative activities.

**Enernet Case Study**

In Ireland, Intel Labs Europe have worked together with a leading electrical heating company Glen Dimplex, the National Grid, Utility Suppliers, and home owners to co-innovate a new electrical energy operating model which will optimally take advantage of renewable energies, new technologies, and maximize efficiencies while lowering costs for all involved. This initiative called the Enernet, which involves integrating electrical grid and Internet technologies is developing rapidly and at the time of writing was being tested where smart electric thermal storage systems are installed in twelve hundred and fifty homes in Ireland, Germany, and Latvia funded by a Horizon 2020 project called Real Value. In the Enernet, demand-side management allows energy users of all kinds to act as virtual power plants. The project involved leading executives from the key players in the Irish energy system who coalesced around a shared vision of developing a future grid in Ireland leveraging digital and other disruptive technologies such as advanced thermal storage technologies. Key executives from all stakeholders used Osterwalder's business model canvas to agree how to configure the solution so that shared value was created for all stakeholders.

A good example of a Horizon 2020 project, which involved all parts of the quadruple helix innovation, is the Enernet—Real Value project. Real Value is a game-changing €15.5m energy storage project funded by Horizon 2020. Real Value commenced in June 2015 and involves installing thousands of Smart Electric Thermal Storage Systems (SETS) into 1250 homes across Ireland, Germany, and Latvia. The Real Value consortium is a microcosm of an ecosystem, which represents the entire electricity supply chain and has secured key expertise from various specialisms including industry, energy services, network, and research organizations with partners from Ireland, the UK, Germany, Latvia, and Finland. A key enabling technology is Smart Electrical Thermal Storage (SETS) developed by Glen Dimplex in Ireland. This technology has been developed to meet householders' space and water heating needs in a low cost and energy-efficient manner while also enabling the electricity industry to exploit its energy storage capacity.

## 8.3 From Clusters to Open Innovation Ecosystems

To enable the establishment of common vision and to create co-creative culture across all stakeholders, including the end-users, open innovation ecosystems need to be fostered. In these, the mash-up and experimentation can create the creative commons needed for the growth of intellectual capital in the ecosystem. Following the research of Lin and Edvinsson, there are clear indications that intellectual capital and especially structural intellectual capital drives competitiveness and innovation. This means in turn that from innovation policy perspective the *interaction fluidity* is a critical feature of any successful innovation system. Fluidity in this context means frictionless interaction, experimentation in real world, and a lot of unexpected, unplanned collisions of ideas, problems, and of course competencies to collide, giving the spark. It is not only about single excellent components in the system, it is centrally about collisions and connectivity.

Previously, it has been shown that the diversity of research teams increases significantly the probability of breakthroughs. Cross-fertilization of ideas is nothing new as such, but what ecosystem thinking does is embed diversity and serendipity in the innovation process more systematically than ever before.

It is important to move from clusters to ecosystems in our innovation system design. There is nothing wrong with clusters, but they tend to be rather monolithic focusing on one sector only. Of course, the clusters reinforce the sector they work in, but the tendency is more towards improving, extrapolating than to create something new. Hence, the emphasis on modern innovation systems need to be increasingly on the 'in-between' areas where creation of new is likely, and as consequence also the fast growth.

## 8.3.1   Living Labs

According to the European Network of Living Labs (EnoLL), Living Labs (LLs) are defined as user-centred, open innovation ecosystems based a systematic user co-creation approach, integrating research and innovation processes in real-life communities and settings. LLs are both practice-driven organizations that facilitate and foster open, collaborative innovation, in real-life environments and arenas where both open innovation and user innovation processes can be studied, experimented with, and spaces/sites where new solutions can be developed. LLs operate as intermediaries amongst citizens, research organizations, companies, cities, and regions for joint value co-creation, rapid prototyping, or validation to scale up innovation and businesses. LLs have core elements but multiple different implementations and manifestations.

The industrially led think-tank for Living Labs strategy in Europe was established in liaison with European Commission, DG Information Society in 2003 to conceptualize the European approach. It soon became evident that the European approach should be focusing on creation of innovation hubs which would build on the quadruple helix innovation model, i.e. strong and seamless interaction of the industry, public sector, research institutions, and universities and finally also the 'people'.

The target was to create attractive environments, which would be attractive for industrial and research investment due to better innovation dynamics. This dynamic would be supported by the public sector and one of the focus areas would be public sector services, which could be co-developed with the user communities, in real-world settings. Part of this thinking was based on the idea to stretch the boundaries of societal behaviour as well, as we saw the connectivity and ICT shared environments (with emerging social media) to change society as well. The quest was to push the boundaries with real-world projects including strong technological development too. Only by doing the research and development with citizens involved, we could see what finally would be acceptable and thus scalable to products and services.

This led to the first concept of Living Lab in European context; a real-world site, not an extension of a laboratory. Important was also the scale as it was seen that for scalability we needed the 'sample users' to be large enough, at least in hundreds (Fig. 8.5).

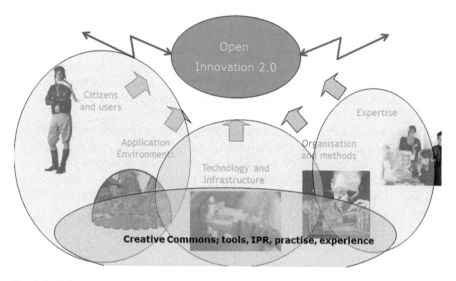

**Fig. 8.5**  Living Labs in a European context

In the figure above, we have all the components needed for European Living Labs: Citizens, application environments, technology infrastructure, organizations, and experts. Important to see is the later addition of societal capital into the picture as functioning Living Labs build strongly on the idea of spillover effects back to the society, giving motivation for all of the stakeholders, including citizens to contribute to the common goal, making the Living Lab a winning game for everyone.

Based on these conceptual thoughts, the European Commission and the Finnish EU presidency launched in 2006 the first wave of European Living Labs, which built a network, European Network of Living Labs, which became later the ENoLL movement. From the first wave the network grew fast under the subsequent EU presidencies to the substantial scale, it has now 340 sites even beyond European borders. And, the network is still growing. What we can say that the Living Labs have now a strong foothold in all European regions, and is being applied as important component in regional innovation systems too.

**IPPOSI Case Study**

An early example of the quadruple helix innovation is the Irish Platform for Patients' Organisations, Science and Industry (IPPOSI) which celebrated its 10-year anniversary in 2016 was established to provide a platform for discussion between patients' organizations, scientists, and industry (and where possible with State Agencies) in Ireland on policy, legislation, and regulation around the development of new medicines, products, devices, and diagnostics for unmet medical needs. IPPOSI is a unique partnership of patient groups/

medical charities, science, and industry. The IPPOSI shared vision is one where state-of-the-art innovations in healthcare are available to Irish patients at the earliest stages. Through expertise, dialogue, consensus building, networking, and influencing they aim to smooth the path in Ireland for new medicines and therapies to move from basic science in laboratories to the patients who need them.

Membership is open to all constituent groups with an interest in healthcare research and development; Patient Representatives, Academic, Science or the Healthcare Industry. IPPOSI is funded by membership fees from industry partners matched by government funds, through the Irish Health Research Board. This allows patient organization/charity and academic partners to join without incurring financial burdens.

In the same vein is the 'Carecity' initiative set-up as a joint collaboration between the NHS in the UK and the Barking and Dagenham council in London, involving the community and patients to help introduce a roadmap of innovations into the council area, which would help improve public and patient health. In a deprived area, the idea is that the community can help co-suggest and co-innovate the solutions in a response to the problems on the ground.

Another example of Triple Helix innovation is Intel's network of Exascale Computing labs which have been established in France, Belgium, Germany, and Spain in conjunction with various European Universities and National Agencies to jointly perform the research which will inform the design of the Exascale computer of the future as well as understanding how best to take advantage of Exascale capabilities. Exascale is seen as the next moon-shot for the computing industry and refers to computing systems capable of at least one exaFLOPS, or a billion billion calculations per second. Such capacity represents a thousand-fold increase over the first petascale computer that came into operation in 2008.

In the creation of the Living Labs in Europe as new technology policy component in early 2000s, the user component was present. The idea was to create hubs of innovation, which would be attractive for industrial and intellectual investment due to faster prototyping and experimentation in real-world settings. In this European Network of Living Labs, the quadruple helix wording was highlighted as the co-creative approach from the first beginning. Independently, other scholars such as (Carayannis and Campbell 2009) have expanded the triple helix concept to include user-led innovation and described this as a quadruple helix innovation process. The first generation of Living Labs were also founded in 2006 with significant EU funding to bring the culture of co-creativity, user-centricity, and quadruple helix approach into wider use in Europe.

Intel's collaborative research institutes and activities with the cities of London and Dublin explicitly comprehend user collaboration and discovery, both at design and run time, and are good examples of quadruple helix innovation. When diverse stakeholders align and combine creative and productive forces, everyone has the opportunity to accelerate and capture the value created.


## 8.4   Collaborative Architecture

A crucial choice is the collaborative architecture and four archetypes have been identified by Pisano and Verganti (2008) which are derived from the combination of modalities of network participation and governance structure (Fig. 8.6). Pisano and Verganti (2008) articulate ways to establish '*which kind of collaboration is right for you*' and they make a bold statement that '*the new leaders in innovation will be those who figure out the best way to leverage a network of outsiders*'.

Key questions include how closed or open should your organizations network of collaborators be and which kind of network configuration and which problems or opportunities should the network tackle. Innovation openness is not a binary variable and at least three dimensions need to be considered (Brunswicker and Hutschek 2010), IP and appropriability strategies, Innovation Search and Sources, and Relationships and networks. Pisano and Verganti (2008) describe four kinds of open innovation configurations: a closed and hierarchical network (an *elite circle*), an open and hierarchical network (an *innovation mall*), an open and flat network (an *innovation community*), and a closed and flat network (a *consortium*). Choosing which network configuration is a trade-off considering all the benefits, costs, and risks of each configuration. Importantly Brunswicker and Vanhaverbeke (2015) identify that both a demand-driven and a widely diversified search strategy can improve the success of small and medium sized companies in launching innovations. Thus, it is vital high expectation entrepreneurs figure out how to take advantage of this new OI2.0 paradigm.

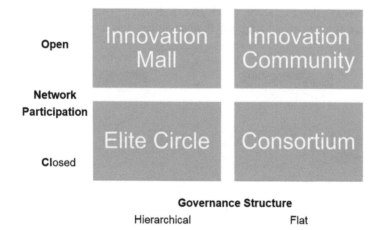

**Fig. 8.6** Collaborative Architecture, Pisano and Verganti

## 8.5  Participative Architecture and Governance

Deciding whom and providing the right incentives to ecosystem participation is vital to success. The Innovation Value Institute is an example of an ecosystem, which was explicitly designed and then orchestrated for success. Six different categories of organizations were brought in to establish a critical mass of ecosystem participants with the right variety of perspectives and expertise. Participants were motivated to participate by a compelling shared vision of driving a structural change in the way organizations achieved value from information technology and were incentivized by the fact that they could generate lots of shared value through using the output of the IVI research either as an end-user or as a consulting organization. BCG, one of the steering patrons of the IVI have used the output with more than two hundred clients (Fig. 8.7).

## 8.6  Business Model Innovation

Business model innovation is a key area of interest and energy currently. According to Doblin's research, it is an area of high return. According to Osterwalder, 'A business model explains the rationale of how an organization creates, delivers and captures value'. In another part of this book, we refer to the concept of design patterns and Osterwalder's business model canvas is a great example of a design pattern. Osterwalder and Pigneur's book on Business Model Generation (2010) is also a great example of co-creation with the book co-created by 470 practitioners from 45 countries.

An important part of driving innovation forward is the creation of a taxonomy, where a taxonomy involves identifying, naming, describing, and classifying the core elements of a business model. Osterwalder in addition presents the business model canvas as a design pattern, which gives a shared language and vocabulary for describing, assessing, visualizing, and innovating business models.

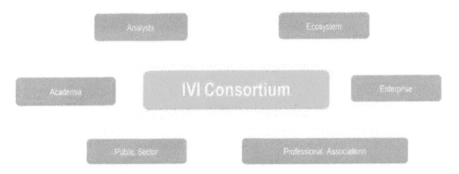

**Fig. 8.7** IVI Ecosystem—Categories of participants

Venkatraman, a strategic management and information technology professor from MIT Sloan School of Management and Boston University, describes the pattern shift from the business-driven era to the innovation-driven era, 'The Five levels of IT enabled Business Transformation'. This theory although very academic and speculative at the time does present the business model change due to ICT in a very straightforward and realistic way.

The operative phrase today is 'IT changes the way we do Business'. The distinctive role of IT is well known in shaping tomorrow's business operations, but when does it become a new value proposition instead of just digitizing the old existing ways of operations?

The first level for leveraging IT functionality within a business typically occurs when managers respond to operational problems or challenges; this causes minimal changes to the business processes and underleverages IT's potential capabilities failing to provide the organization with more advantages (Fig. 8.8). Because the decision is based on need and no radical changes are made to the business processes, competitors are easily able to imitate such practice. The decision should be motivated by a focus on differentiation and strategic effectiveness.

The second level of integration occurs when there is business process interdependence and technical interconnectivity, however as well as in the previous level; the decision should be based on improving a specific efficiency and not simply the result of automating already inefficient business processes.

The third level is business model redesign by altering 'First Principles'. The benefits from IT functionality cannot be fully accomplished if superimposed on the current business models, just like the two previous levels. That is because the current business processes come from a set of organizational principles from the industrial revolution. ICT functionality has altered some of these 'first principles'; ICT should be used as a lever for designing the new organization and associated business processes.

The fourth level is based in the previous level, altering the 'first principles' and in addition considering that the exchange of information amongst participants in

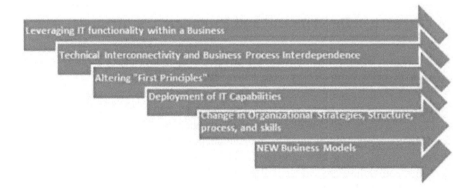

**Fig. 8.8**  The Five Levels for Business Model Change, with a twist

a business network by using IT capabilities is not only effective and efficient for all participants, but also allows for further innovation and development of enterprise by using the innovations and developments from other enterprises to the bigger benefit of all.

The fifth level is considering that the business networking benefits and the role of information technology have direct impacts on the logic of business scope and the resulting redistribution of revenue and profit, meaning that, tasks may be eliminated; some tasks may be restructured, and some tasks may be expanded. IT's potential benefits are directly connected to the degree of change in organizational practices, strategies, structure, processes, and skills. The sixth level is not part of the original five levels of business model change because it is the NEW business model.

### 8.6.1  Business Model Experimentation

Business model experimentation is about doing things that are unfamiliar and different but present a great opportunity for business model innovation. Business model experimentation considers the facts and assumptions that are already known and trialing others that might be risky, but in business model experimentation tools to reduce these uncertainties are used to decrease risk and increase success rate.

One of the best business model experimentation concepts in the industry is the business model canvas introduced by Alexander Osterwalder. His model considers all basic aspects of business models and allows for experimentation and flexibility; it accepts modifications on-the-go and opens the business for different opportunities in regard to customer segments, solutions and benefits, value and cost levels, pricing, differentiation, channels and relationships, as well as resources and partnerships. The business model canvas consists of cost structure, key resources, key partners, key activities, customer relationships, customer segments, value propositions, channels, and finally revenue streams.

## 8.7  Ecosystem Development and Evolution

Bruce Tuckman's stages of team or group development model can be applied to the development of evolution of an ecosystem. Tuckman's (1965) four-stage model describes the four stages a team will go through before the team becomes high performing. In the initial stage 'forming', the team meet and agree what are the goals and challenges. Typically, team members act relatively independently and the leader has to take significant responsibility for direction and guidance. In the second stage of development 'storming', the leader often has to take the role of a coach as trust is not yet established across the team and team members' view for position as they try to establish their role and relationships within the team. 'Norming' is the third stage of development where consensus and agreement emerge as the dominant decision-making mode and the leader acts as a facilitator and enables. In the third stage, individuals often move from a compliance mentality to a commitment mentality. The fourth phase is the 'performing' phase where the members of the team are aligned around and execute to a shared vision. In this mode, the team often commit to over-achievement goals and operate with a lot of autonomy. A key distinguishing factor of a high 'performing' team is trust, while disagreements might well occur they are resolved positively within the team. A high performing team delivers synergy where the collective output of the team is much higher than the sum of the contributions of the individuals on the team.

For an ecosystem a keystone player, as defined by Michael Moore is needed to act as an orchestrator and to determine the beat rate of the ecosystem. When an ecosystem gets to stage four, it is high performing and individual members acting autonomously almost like natural occurring biological ecosystems with individual entities efforts and contributions aligning and accelerating the collective impact of the ecosystem. This is not to say that there is no competition amongst members of the ecosystem; a healthy 'coopetition' exists and the right balance between collaboration and competition is achieved when ecosystem members adopt a 'win-more, win-more' mentality.

Ecosystems may well organically evolve but where the ecosystem is explicitly orchestrated and managed this significantly increases the chance of success and the efficiency and effectiveness of the ecosystem as well as maximizing the sustainability and longevity of the ecosystem.

## 8.8   European Innovation Ecosystem and Scoreboard

The annual innovation union scoreboard provides a comparative assessment of the research and innovation performance of the EU28 member states and the relative strengths and weaknesses of their research and innovation systems. It helps member states assess areas in which they need to concentrate their efforts in order to boost their innovation performance. This is a crucial tool to help evolve national innovation ecosystems and system by pinpointing the areas in need of most improvement. The EIS is holistic in that it looks at input, output, and intermediates innovation indicators (Fig. 8.9).

Ireland is a great case study of how to use the EIS. Ireland began a focus on research and development and initially from a very low base had a focus on generating research publications. In 2008, Ireland had three times the average volume of scientific publications but was only at the European average in terms of top ten percent of scientific co-publications thus indicating that quantity was delivered over quality. With the appointment of a new Director of Science Foundation Ireland Mark Ferguson, a focus on quality research with impact worked to improve quality over quantity (Fig. 8.10).

Additionally, new tools are emerging to help manage and get better yield from pan-European R&D investments. One such tool is the innovation radar work led by

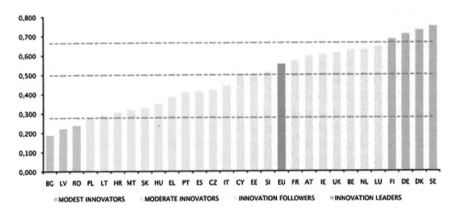

**Fig. 8.9** Innovation Union Scoreboard

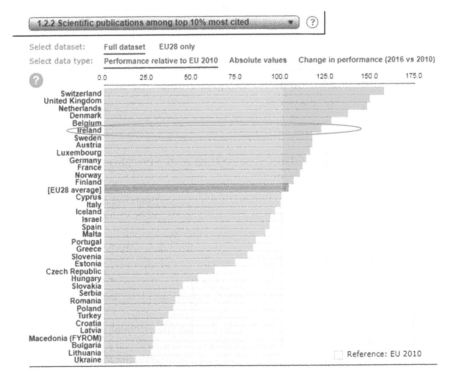

**Fig. 8.10** Health of Ireland Innovation System vs. EU Average

Eoghan O'Neill in the European Commission, which aims to systematically identify the winning innovations and innovators, which could be scaled. Additionally, it will help identify the innovation results and the differentiators for success in EU projects to assist in selecting future projects. Thus, innovation radar results will also be used in designing new action types for the EU research and innovation programmes and to highlight the impact of the investments. Also we highlight the need of new professions in these open innovation ecosystems; the curator taking care of quality thematic contents, the bridger making the connections between curated contents and ideas, the orchestrator driving the talent together for a common objective, and finally the systems designer who makes all this possible from process and connectivity perspective.

**Case Study: Cambridge Wireless**

Cambridge wireless is a contemporary example of ecosystem orchestration and stewardship. CW is an organization, which is managing a vibrant community with a rapidly expanding network of nearly 400 companies across the globe interested in the development and application of wireless and mobile technologies to solve business problems. CW connects those companies and stimulates collaborative innovation through a range of thought-provoking high-profile networking events.

In addition to these high-profile VIP networking activities, CW run nineteen special interest groups (SIGs), each focused on a specific technology and/or business area. SIG meetings provide opportunities for member organizations to meet, learn from each other, and explore opportunities to work together.

CW's interest is in practical solutions to business needs in the areas of mobile, networks, Internet, semiconductor, advanced software, and particularly anything cellular. While the community is global, it does have a geographic locus in Cambridge and helps turbocharge the local ecosystem.

CW's mantra is to network, collaborate, and innovate representing a very valid innovation pathway. CW is a not-for-profit organization that is owned by its members, with a governing board that is elected by the membership. Members are drawn from all parts of the wireless-enabled world, from securely connected devices, networks, smart phones, software, and applications, through to data analytics, content delivery, telecommunications, and satellites.

## 8.9   The Entrepreneurial State

Mariana Mazzucato in her book 'the entrepreneurial state' says that the government needs to do more than create the right conditions for innovation such as policies that favour industrial growth and better education. She argues that the government needs to initiate large-scale innovation initiatives that are society transforming. This is at the essence of quadruple helix innovation where a shared vision is established with a major transformation goal. Often the vision may come from outside the government or public sector but the government needs to adopt and intellectually and financially support the innovation program that strives to implement the vision.

Professor Mazzucato emphasized the important role of the state and found after a detailed review of case studies that it was only after an entrepreneurial state had made a high-risk investment that the private sector found the courage to invest. Her ground-breaking research showed that many of the breakthroughs required by Apple to develop the iPhone were funded by government research in the USA and Europe.

## 8.10   Policy: National Innovation Strategy

The role of the government is changing from not just enabling innovation and creating the conditions for innovation but also to be an innovation driver setting key focus areas or smart specialization strategies and actively participating in the innovation process.

According to the European Innovation scorecard, Ireland is an innovation follower but the Irish government and the department of jobs, enterprise, and innovation have defined a national innovation strategy which sets out an ambition and a strategy to make Ireland an innovation leader in a number of defined areas—this is called Innovation 2020 and sets out a 5-year strategy and a roadmap towards a goal of making Ireland an innovation leader towards creating a strong sustainable economy and a better society. Ireland in recent years create a research prioritization process to focus the countries innovation capacity on efforts on a limited number of key high potential focus areas. Having completed and implemented the research prioritization exercise, the next step in maturity was to implement an innovation strategy. This is in line with the aphorism 'Research turns money into knowledge, Innovation turns knowledge into money'.

According to Innovation 2020, 'Innovation—the quest to find solutions that are original, more effective and deliver positive change—is at the heart of Government policies for enterprise, education, social and cultural development, and the delivery of public services'. Interestingly, Ireland has one of the most open economies in the world and has attracted through a 'low tax and talent' policy a disproportionate percentage of US foreign direct investment in Europe into Ireland. Through sustained investment, Ireland has been transformed from a primarily agrarian-based economy to a leading centre for education, research, and industry across a range of cutting-edge disciplines. From this agrarian base, Ireland today has nine of the top ten global pharma and ICT companies with European headquarters or significant presence in the country demonstrating the benefit of sustained innovation in innovation policy.

**The Case of Ontario: Incentives Matter**
The state of Ontario, which had lagging economic performance compared to peers states and following the creation of a taskforce[1] identified that Ontario needed to encourage innovation to compete in the modern global economy. A core conclusion of the report was that innovation plays a vital role in the

[1] Finding Its Own Way: Ontario needs to take a new tack, Institute for Competiveness and Prosperity, Ontario (2014).

state's prosperity but identified that Ontario lagged its peers in key measures of innovation such as R&D expenditure, commercialization of innovations, and patent output.

After a detailed analysis, the taskforce came to a very interesting conclusion that it was not the amount of money that was spent on innovation that mattered but the incentives for innovation that mattered. They found that indirect support and a patchwork of different support programs failed to generate major chance. The taskforce concluded that Ontario needed to provide more direct support for innovation, particularly for large businesses. Three action items were agreed.

- Increase the amount of business R&D expenditure
- Introduce policies which encourage the commercialization of R&D and give better protection to IP
- Broaden the understanding of innovation and incorporate innovation education into Ontario's education curriculum

## 8.11 Co-creation

Ramaswamy's and Goiullart's book (2010) on the power of co-creation and how to boost growth, productivity, and profits nicely conceptualizes a growing trend in both business and societal co-innovation. Central to the concept of co-creation is a shared vision of how two organizations have a shared vision of how they could and do create shared value. Intel Labs Europe collaboratory (co-labs) with SAP, BT, and Nokia were excellent examples of this where companies with complementary strengths, capabilities, and interests can align and accelerate innovations to the market.

Increasingly, open flexible functional platforms for rapid evolution of solutions are important to provide platforms for innovation. Alexander von Gobain, former Chair of the European Institute of Innovation and Technology, described the EIT Knowledge and Innovation Communities (KIC) as 'Innovation Factories' where each KIC tries to establish a distributed ecosystem to support innovation and entrepreneurship in a specific domain.

One example of an open functional innovation platform is the GE ecomagination challenge for which 3800 ideas were submitted many by entrepreneurs with strong track records. The $200m funding commitment by GE and four venture capital partners is a very attractive way of funding would-be entrepreneurs and the linking of similar ideas and the ability to comment on other people's ideas create a platform for rapid emergence and evolution. The ecomagination program is another great example of share value at work representing GE's commitment to imagine and build innovative solutions to environmental challenges while driving economic growth.

The Ireland-based Innovation Value Institute has developed a global platform for research and innovation in creating tools to help Chief Information officers (CIOs) better manage their IT capabilities and improve business results through IT. Most participants in the IVI consortium both contribute knowledge and use the collective knowledge generated. Shared value is not only created through adoption of the research outputs called 'artefacts', but the IVI is also catalysing the formation of new start-up consulting firms as well as helping existing consulting firms scale through provision of assessments and services. All the global members of IVI sign up for the IVI vision, which is to drive a structural change in the way companies and governments get value from information technology.

## 8.12  Visualizing Innovation Ecosystems

Innovation ecosystems are difficult to manage and characterize. Increasingly, there are software tools becoming available to help map and graph what is happening in the ecosystems to help inform executives about future decisions. In a rapidly evolving ecosystem such as that for the Internet of Things where there is no keystone player and no dominant design this can be very helpful. Equally, for a start-up the StartUp Genome ecosystems report, which analyses the vibrancy and leadership of different cities for their start-up culture and environment, can be invaluable. StartUp Genome apply science to ecosystem measurement and visualization and have developed a four-stage model to describe which part of the ecosystem life cycle a particular city or ecosystem is in.

Activation describes the start-up phase of an ecosystem where there is a low number of start-ups, typically less than a thousand and where there are resource gaps. In the globalized phase, there have been several high profile and large exits (>100m) which have put the ecosystem on a global map. In the expansion phase, there are more than two thousand start-ups and there have been a number of multimillion dollar exits and unicorns, which have elevated the status of the ecosystem on a global level. The integration phase is the pinnacle of the models where the size and results of the ecosystem are world-leading as measured by rate of early-stage success, number of unicorns, and rate of exits.

StartUp Genome LLC ranks cities annually using criteria such as performance, funding, reach, talent, and start-up experience with ecosystems such as Silicon Valley, New York, London, Beijing, and Boston being the top five start-up ecosystems in the 2017 report. Start-up genome make their database of data available for a small fee to facilitate further research and analysis by different stakeholders perhaps looking for the right place to locate their start-up or a local authority figuring out what gaps should be closed to make their ecosystem more competitive and attractive.

## 8.13  Intel Labs Europe Ecosystem Management

By 2016, Intel Labs Europe (ILE) numbered more than fifty labs across Europe. There is a lot that can be achieved with this number of labs. However, the real power of ILE was the number of organizations that interacted with it in its ecosystem

**Fig. 8.11**   Intel Labs Europe Ecosystem

leveraging OI2 principles. In 2016, there were more than seven hundred organizations participating in joint research with ILE as depicted in the ecosystem diagram above (Fig. 8.11).

Alignment was achieved by sharing ILE's Innovation agenda, which had a vision of a Digital Europe, using digital technology to transform societal level systems in Europe through digital technologies. This became a shared vision, which extended over Intel's research participation in European programs. With ILE labs participating in over seventy EU framework seven project, ILE was able to achieve influence and results far beyond its own footprint and investments in Europe. Additionally, it helped to build generic innovation capacity with researcher and developers who developed an aptitude and affinity for using Intel Architecture in projects.

## 8.14   Makers and a New Innovation Ecosystem

Innovation ecosystems can be physical or virtual, encompassing all stakeholders driving for common goals. Ideally, ecosystems have hundreds of projects/experiments simultaneously, cross-fertilizing each other.

Engagement platforms are important to test, verify, and develop the ideas to innovation. The maker movement is an umbrella term for independent inventors, designers, and technology tinkerers. The creation of FabLab type of facilities and maker spaces is making experimentation and prototyping accessible and affordable for all stakeholders.

---

**Arduino Case Study**

Arduino is an open-source hardware and software prototyping platform intended for both professionals and hobbyists who are interested in creating interactive objects or environments. The platform is extendable by using 'Shield' add-on boards. Originally invented in Italy, the open-source Arduino movement has become a favourite of makers worldwide with its low-cost and usability helping democratize the ability to use digital technology for innovation by a global maker community. The Arduino's community project products are distributed as open-source hardware and software, which are licensed under the GNU General Public license permitting the manufacturing of Arduino board and software distribution by anyone. It was estimated that by 2013, there were over 700,000 Arduino boards in use across the world and the growth and value of Arduino and its associated community is an excellent example of OI2 at work.

Beginning in 2003, the Arduino project started as a program for students at the Interaction Design Institute in Ivrea, Italy aiming to provide a low cost and simple way for both professionals and beginners to create solutions using sensors and actuators such as thermotstats, motion detectors, cameras, and even robots. The name Arduino comes from a pub in Ivrea where the founders used to meet. This example demonstrates the power of a shared vision to create a worldwide movement with a report greater than a hundred million visits annually to the Arduino site by 2017. Also in 2017, the Arduino founding team were awarded a European Innovation Luminary award by the EU OIPSG and associated academy.

---

## 8.15   Innovation Ecosystems Orchestration

In the past innovation, ecosystems grew organically and took time to emerge. Today, there is a recognition that innovation ecosystems can be designed and orchestrated for success. Innovation ecosystems can exist in many forms, in a geographic region, across a particular industry or increasingly they can be virtual not tied to a specific geography. Geographical or virtual ecosystems can span or traverse a number of business ecosystems. Bill Aulet of the MIT Entrepreneurship Centre describes seven key pods that make up the innovation ecosystems: government, demand, invention/innovation, funding, infrastructure, culture and, most importantly, entrepreneurs themselves.

Rather than letting innovation ecosystems evolve organically, there is increasing recognition of the need to explicitly invest in and manage national and regional innovation ecosystems. For example, to spur economic development in the state of

Connecticut, Governor Daniel Malloy as part of the state's innovation ecosystem program announced four innovation hubs in October 2012 hosting financial, technical, and professional resource offerings for business. Initially costing $5 million they were funded through the Malloy's 2011 Jobs Bill. Another example is the 2010 Irish Innovation Taskforce report commissioned by the Irish Prime Minister, which put developing and supporting the national innovation ecosystem at the centre of the national strategy for innovation 'Building Ireland's Smart Economy'.

However, innovation ecosystems are extending beyond regional and national boundaries. Victor Hwang (2012) identified the hugely important impact of IT and networks when he wrote, 'Fortunately, the social networks of the future no longer need to be imprisoned by geography, as we can build overlapping social networks that traverse the boundaries of the past'.

It is becoming apparent that a new more powerful kind of open innovation is happening where we have broad collaboration across many actors, with the locus of competition not only revolving around particular companies but around competing ecosystems. Consider the case of launch of Microsoft's Surface tablet, which at launch was targeted to have five thousand apps available for consumers to use. At first blush, that might sound impressive, but compared to the seven hundred thousand, which were already available on the Apple iPad (Wall Street Journal, October 24, 2012) it puts in perspective the power of a developer ecosystem surrounding a platform.

The analogy of linear momentum can be applied to innovation ecosystems with momentum being mass by velocity. The ecosystem, which has the most mass (either value chain participants or users) and the highest velocity, is in the strongest position to win. However that is not always the case and small ecosystems, which are agile, and have high velocity can overcome the inertia of an established ecosystem. Also past success is not a guarantee of future success. Witness the struggles that Nokia/Microsoft are encountering in try to compete with the Apple iPhone and Android ecosystems.

Would-be entrepreneurs who attach themselves to fast-moving ecosystems stand to derive significant benefits from the velocity of an ecosystem. We must not confuse speed and velocity. Velocity is speed by direction, and where innovations are delivered in the context of a vision vector that is much more likely to have longevity rather than ad hoc innovations that are delivered with speed. As Stephen Covey says, 'begin with the end in Mind'.

And yet many successful entrepreneurs did not have an end state in mind and their products or services evolved as they learnt more and collaborated with partners. With increasing interconnectivity, the Internet itself creates a platform for emergence, where half-baked ideas and prototypes can be quickly piloted and iterated to deliver new value products or services. Indeed, with the continuing advance of information technology, many products are becoming much more IT intensive and this creates new opportunities for technology-based entrepreneurs.

One example of new innovation ecosystems established is the GENIVI nonprofit industry alliance committed to driving the broad adoption of an In-Vehicle Infotainment (IVI) open-source development platform. The mission of the GENIVI consortium is to drive the broad adoption of an open-source development platform by aligning automotive original equipment manufacturers (OEMs) requirements and delivering specifications, reference implementations, and certification programs that form a consistent basis for an innovative infotainment system. GENIVI was formed by a broad consortium of players such as BMW, Jaguar LandRover, and Intel to drive efficiencies and innovation across the automotive ecosystem for the so-called in-car infotainment systems.

Rather than each automotive company and supplier, having to build custom hardware and software the goal was to develop a standard platform for infotainment that companies could then innovate on top of. The fact that many competing automotive companies and suppliers could come together successfully to collaborate to lower costs, increase agility, and accelerate innovations in in-car infotainment is a great example that something is fundamentally changing in the nature of innovation and collaboration.

Another similar example from the computer networking industry is the Greentouch consortium led by Alcatel Lucent which is a consortium of competitors and suppliers working together around a unified mission of improving the energy efficiency of networks by a factor of one thousand from current levels.

Jeff Alex at SRI uses the metaphor of a biological ecosystem to describe the nature of a business ecosystem. A key characteristic of an ecosystem is that it is evolving with organic, diverse, and symbiotic attributes. The principle of synergy is central—the idea that through collaboration entities can deliver something, which is unattainable on one's own. Ecosystems are also complex-adaptive systems. Once the raw materials are put in place and the initial relationships and couplings established they are often self-organizing and self-regulating according to Darwinian principles.

The longevity of an ecosystem and its relative fragility or resilience is dependent on whether it is organized as a value chain or as a value cycle. This is a critical distinction. A value chain is built for linear production where 'you use it (products, resources) up, throw it away and not worry about where it goes when it's gone, because that is someone else's problem' (Haque, 2011). 'Linear production is built to make stuff that dies after a fixed life cycle, over and over again'. Now how could that be sustainable?

In a value cycle (Haque, 2011), resources are used intensively without depleting them. In fact, the more intensely and durably that resources can be recycled or at least exchanged the more longevity is assured as prices drop because each cycle amortizes and offsets the capital and expenses, costs of production like factories, people, and knowledge. In a value cycle scenario, an ecosystem can approach the efficient frontier (Marjanovic et al. 2012) because more value is created over a longer period of time with less risk.

Ultimately, the goal is to have innovation ecosystems, which are autocatalytic—where a reaction or ecosystem is catalysed by a product developed by the ecosystem. Again the Apple App store ecosystem is an example of this where new apps added help catalyse the creation and adoption of further apps.

A key task of the ecosystem orchestrator is to achieve the right balance between control and enabling/inspiring. The most effective ecosystems will be those that are empowered by a shared vision rather than controlled. When there is commitment rather than compliance, the results will ultimately be much stronger.

# Chapter 9
# Designing for Adoption

*'Innovation is not innovators innovating but customers adopting'*

Michael Schrage, MIT

A hallmark of OI2 is the focus on adoption and innovation in OI2 is defined by the creation and adoption of something new, which creates value for the entities that adopt it. Similar to the movement of 'Design for Manufacturing' where engineers designing products consider how to make products more easily manufacturable, designing and innovation for adoption is critical for successful adoption of innovations. According to the OECD, 80% of the value of innovation comes from the successful adoption of an innovation with just 20% of the value coming from the creation activity. This is of course obvious and often we consider the hard part of innovation to be the creation phase and this is where most resources are often committed. However ironically as stated most of the value from an innovation comes when it is adopted which is the phase that is least considered or invested in. By carefully considering the factors required for successful adoption, we can significantly increase the probability of successful adoption. We see that the quadruple helix co-creation process is crucial for rapid scalability of new extended products, especially in new markets.

Users are an enormous source of innovation and when products or services are co-created they are designed for adoption from the get-go. Erik von Hippel's research shows that the majority of significant innovations in the semiconductor industry over a 30-year period came from lead users. Co-innovation with users or direct innovation led by users is especially powerful. Designing a platform where users can innovate on a foundational set of assets with an associated set of standards can be very powerful. The Apple app store and platform is powered by an enormous community of app developers whose imagination and energies fuel the increasing adoption and use of the Apple platform.

Usability in OI2 can be achieved faster by collaborating with users with real-world experiments in living labs; here, we have the notion of fail fast, learn fast, and scale fast. Experimentation using agile developments is crucial in evolving utility, user experience, and usability to meet the needs of the user community.

The six components in the 6U adoption pattern are

© Springer International Publishing Switzerland 2018
M. Curley, B. Salmelin, *Open Innovation 2.0*, Innovation, Technology, and Knowledge Management, DOI 10.1007/978-3-319-62878-3_9

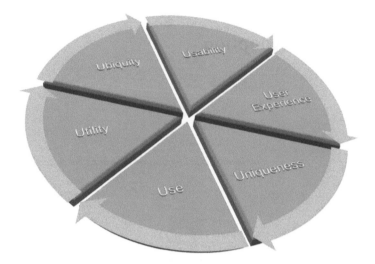

**Fig. 9.1**  6U Designing for Adoption Pattern

- Utility
- Usability
- User experience
- Users
- Ubiquity
- Uniqueness

When these attributes are sought, especially with the involvement of lead users using a platform-based innovation approach, the probability of subsequent adoption is substantially higher (Fig. 9.1).

## 9.1  Utility

By utility, we mean the quality or state of being useful and we explore what is the value or usefulness does the innovation provides? perhaps doing something better, faster, or cheaper than before or else creation of a whole new function that was not available or possible before. Innovations which are quirky but do not provide utility quickly fade from view. Products and services need to be designed for **Use.** A beautiful product or service, which has an unacceptable mean time between failure (MTBF) or 'blue screen' rate may not have longevity. With the increasing emergence of the sharing economy, innovation design and operation criteria will need to be optimized for utilization and longevity of assets as well as utility.

Industry commons is a good example of utility functions enabling integrated use of agreed functionalities across technologies, in a technology agnostic way. An open approach is essential as only then the industry commons can have a wide coverage across industries.

## 9.2   Uniqueness

Uniqueness is a critical factor for adoption. With Digital, it is increasingly possible to do things that even several years ago appeared impossible. Indeed, it seems the Adidas marketing slogan 'Impossible is nothing' seems increasingly to apply to Digital. When we consider platform innovations such as Airbnb or Uber, the uniqueness was the scale at which they were able to bring new services to the market, making unused capacity in private homes or cars available for use and providing the capability to meet needs of people in real time. Both of these services are exemplars of the kind of innovations, which make up the 'Sharing Economy' where the utilization of assets are improved in everyone's interests. For a digital wallet, the ability to store and automatically exchange a digital coupon at a store and to automatically accumulate loyalty points rather than having to produce paper coupons at a till will provide a unique capability which will speed adoption.

## 9.3   Usability

Usability examines how usable is the new innovation or service. For example, many in the financial services industry are pushing the deployment of digital wallets but for many users the set-up efforts required and lack of merchant infrastructure currently deployed means that credit cards currently provide better usability than current generation digital wallets. Additionally, some have contrasted the implementation effort of current generation digital wallet software at issuing banks to be the equivalent of an ERP (enterprise resource planning) software installation, which significantly increases the barrier to adoption of digital wallets across the broad financial system. Innovations may be unique and may bring much value but if they are complex to use then adoption can be slowed or indeed fail. Co-creation with the users focuses the development process towards adoptable, usable solutions which in the innovation process are verified in real-world settings.

## 9.4   User Experience

User experience is increasingly considered a critical factor in the adoption and use of innovations. Analysts often refer to the experience economy and increasingly those products and services, which provide better user experiences are more quickly adopted and can command a healthy price and margin premium. When a product or service provides both utility and a good user experience, then the probability of adoption is significantly increased. There were many MP3 players in the market before Apple launched the iPod, but it was the user experience of the iPod, both in terms of touch and feel and the backend services provided through iTunes that drive outrageously successful adoption compared to earlier market entrants.

## 9.5    Ubiquity

Innovations that take advantage of the network effect, whereby the value of the innovation increases with each additional user, can have dramatically increased rates of
adoption once a critical mass of users is adopted. Here, innovations take advantage
of network, software, information, silicon capabilities, and economics and designing
carefully to create utility and incentives before an innovation is launched can have
dramatic effects on adoption rates. Platforms such as Facebook, Airbnb, and Google
all exhibit network effects and are very hard to displace once a network effect takes
hold. There are many criteria, which influence adoption of innovations—two important ones are described in the Bass Diffusion equation, which are the innovation
index and the imitation index. However, the imitation index appears to be much more
influential than the novelty index in successful consumer adoptions.

## 9.6    User-Driven Innovation and the Reverse Innovation
         Pyramid

Jean Claude Burgelman was one of the first policy makers to identify the trend of
user-led and user-centred innovation. Burgelman outlined the shift from the user as a
research object to the user as a research contributor to ultimately the user as a full
research participant. Erik von Hippel argues that innovation is now being democra-
tized in that users, supported by ever improving computers and communications,
have ever improving ability to develop their own services and products. Von Hippel
also noted that these users often freely share their innovations with others creating
new intellectual commons and associated user-driven innovation communities, a key
characteristic of OI2. Intel's joint innovation lab with Nokia in Oulu where the mayor
and the collective citizenry were an integral part of the innovation process is another
good example of this. In Oulu, the municipality provided a good user engagement
environment with open Wi-Fi for all covering the whole city. The university and
research was much focused on mobile communications and their applications.
Finally, the city offered also public services (e.g. health services) to be developed in
real settings, with real citizens.

The Internet is itself a great example of a user-driven phenomenon where once it
was established, without central governance it has continued to grow, evolve, and
deliver more and more utility driven often by the users. Johnny Ryan (2010)
describes the Internet as a 'centrifugal force, user driven and open'.

The Internet also acts as intellectual supercollider bringing people and ideas
together with low collaborative friction and fast feedback loops.

Dell Computer encourages users to submit new product features and ideas to it
and also allows customers vote on the top new features they would like to see in Dell
products. The standout example of user-led innovation is the powerful developer
community—many just individuals who contribute new apps to the Apple iPhone,
which transform it to a modern day digital Swiss army knife.

In parallel, the EU IT advisory group (ISTAG) identified more than a decade ago the future trend of the consumer moving to a prosumer. ISTAG recommended strongly to invest in EAR (Experimentation and Application Research) methodology in European frameworks.

YouTube is an excellent contemporary example of this where users not only consume content but also create and upload content. YouTube and Facebook are forerunners of a potential much deeper level of mass collaboration that is enabled by pervasive high-speed networks and ever-increasing computing power.

The European Internet Foundation in their seminal report 'the Digital World in 2025' identified mass collaboration as the emerging dominant paradigm. *Wikinomics* (Tapscott and Williams 2006) described the positive effects of peer production. The Innovation Value Institute consortium physically based in Ireland although virtually distributed across the world is an excellent example of mass collaboration amongst IT executives in which CIOs have created a model 'built by CIOs, for CIOs'.

With predictions of potentially more than fifteen billion connected devices by 2015, collaboration will not be limited to increased person-to-person collaboration. There will likely be waves of machine-to-machine collaboration and person-to-machine collaborations. It is projected as the car becomes part of the network (Bill Ford announced at Mobile World Congress in 2012 that the car is now part of the network) that future cars may transmit three thousand messages per second to other cars in their vicinity. This will call for a whole new meaning of embedded computing, but will also enable mass synchronization of traffic likely resulting in earlier arrival times and more fuel efficiencies.

## 9.7   Reverse Innovation Pyramid

Geleyn Meijer, Gohar Sargsyan et al. in the 2012 OISPG report on the socioeconomic benefits of open service innovation neatly conceptualized the new aspect of user-led innovation and called this the reverse innovation pyramid as shown in the figure below (Fig. 9.2).

Rather than innovation being 'done' to users, in the new open innovation models, users are participants in the innovation process and there is often shared wealth created. The Apple App store and its massive community of app developers is a standout example of this.

Open Innovation 2.0 enables a reversal of the typical innovation pyramid with the users as a primary source of innovation and indeed the user innovators potentially sharing part of the revenue stream. The Lego Kuuso platform http://lego.cuusoo.com/ mentioned before is a great example of this. The development of the Lego Kuuso platform is also an example of reapplication as Lego piggybacked on the existing Kuuso community in Japan.

When users are intimately involved or indeed drivers of innovation, then adoption is almost guaranteed as these lead users help make sure the innovation actually solves a real problem or helps seize an opportunity.

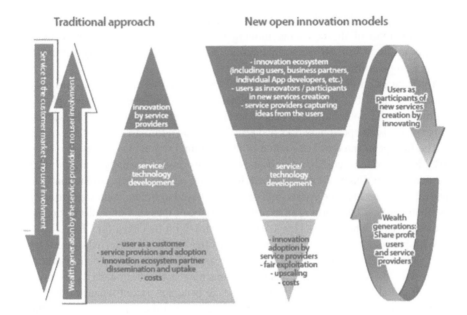

**Fig. 9.2** Reverse Innovation Pyramid: source Meijer & Sarsgyan

## 9.8   The Power of Crowds

Through the power of the crowds, innovators and thinkers have found new ways of generating value to all stakeholders through hedonism and stimulation of the mind as well as by sharing this newly found pleasure in innovation and information sharing.

One of the advantages of the power of the crowds is the positivism that the crowds bring into the equation. Usually, people that find pleasure in the discovery of new ideas and are passionate about innovation are positive individuals, and positivism is a very powerful tool when it comes to generating value. When individuals possess a 'forward' mentality and continue exploring thoughts always thinking that something positive is going to come out of it, more often than not something positive actually comes out of it. Sharing this positive ideas can create a domino effect and make other innovators realize how powerful their positive ideas are getting around and how working and sharing ideas in a positive environment can generate more innovative ideas and create even more value.

The value of Open Innovation 2.0 is seen when not only information but also facilities and services are shared. Developing solutions so that companies can profit from each other in the use of intellectual property, facilities, technologies, and even resources can be cost-efficient, effective, and adds value. Integrated research, development, management, marketing, technology, and other complex tasks could be beneficial.

An example of the role of the crowds in open innovations is that when employees change jobs, they take their knowledge with them, resulting in knowledge flows between firms, and even countries!

The role of the crowds becomes even more important when the acquired knowledge is combined with internal knowledge from the new environment and that external knowledge generates new ideas and subsequently innovation, this is where the *value added* is found.

However, even more value added is found when different crowds get together such as large corporations, small, medium and micro enterprises, universities, technology centres, and associations amongst others. It is natural for people to share ideas and thoughts, but getting people to innovate requires the right crowd to be together.

## 9.9   Crowdfunding

An extension of crowdsourcing is the idea of crowdfunding made popular by platforms such as Kickstarter and Indiegogo. Crowdfunding is a mechanism of venture funding where many people contribute small amounts of capital, typically over the Internet. According to Dan Marom, a crowdfunding pioneer 'Crowdfunding is powerful because it transcends finance; the mechanism is a vehicle for marketing, innovation, market validation, sales, and intrapreneurship'. Crowdfunding can be very effective as it both secures start-up funding and also secures customers based on a compelling share vision articulated as part of the funding campaign. Crowdfunding also helps democratize venture funding by making new funding mechanisms available to people and organizations who otherwise would have great difficulty accessing funding.

## 9.10   Social or Peer Production

Social or peer production is a new form of socioeconomic production in which many people work collaboratively together towards a common goal. Often commons-based peer production projects are designed without a need for financial compensation by others—the example of Linux being a hallmark of a peer production activity.

At the core of peer production are the principles of modularity and integration. Objectives must be divisible into components or modules that can be independently produced and then seamlessly integrated.

**Lego Case Study**

A good example of a win-win situation for all stakeholders is the LEGO innovation community.

LEGO has millions of active users of the toys. From that base they have created an engagement environment where participants to the forum (unlimited, but in practice hundreds of thousands of people worldwide) can come with their LEGO toy ideas for public crowd voting. Lego's vision is to inspire and develop the builders of tomorrow and their Kuuso platform enables children to design new LEGO models and submit them to review a Lego review board. Prior to being approved for review by the Lego review board, they have to get 10,000 votes from the community to pass the review threshold. Users who have their designs approved by the Lego review board receive 1% of the revenue from sales of the design when it is manufactured. For Lego, the benefits are many, including building brand affinity through participation, sourcing new ideas externally and creating new sources for revenue. For the users, Lego have created a platform to help them express their creativity, help develop entrepreneurial skills as well as having fun through play.

## 9.11   Adoption Pattern Analysis

An adoption pattern analysis can be helpful in identifying whether a new innovation is likely to be successfully adopted or not, as shown in (Fig. 9.3). Adoption is assessed against the six core components of the designing for innovation pattern. The adoption pattern analysis shows an estimate of the current mapping of a digital wallet and a future digital wallet compared to a credit card. It is likely that in the future digital wallets will be pervasive but some functions need to be improved upon before the conventional physical wallet/credit card will be displaced.

Users may often compromise on one 6U dimension to take advantage of a particularly strong capability in another 6U adoption factor. For example, older generation mobile phones which did not have large colour screens typically had battery lifetime longer than a week while smartphones often have battery lifetime that are shorter than a day; however, users are willing to compromise on battery lifetime due to superior functionality of the smart phone. Thus, utility and user experience trumps design for use or usability.

### 9.11.1   API Adoption

Since API's will be a scaling engine for the data economy you also have to think about the user experience for developers. Thus well defined and easy to use and easy to access API's are essential to help drive developer adoption. In fact, the six key characteristics for adoption apply equally well and are essential to successful API

## Adoption Pattern Analysis

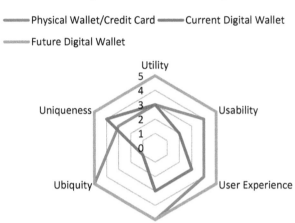

Fig. 9.3   6U Adoption Pattern Analysis

adoption. In future, APIs and their associated documentation will need to be fully machine readable as a cohorts of API users may well be bots.

## 9.12   Adoption Focus: Crossing the Chasm

Geoffrey Moore, building on the seminal work of Everett Rogers, identified the key factors required to drive adoption of new products. Moore identified five kinds of adopters of new products or services, innovators, early adopters, early majority, late majority, and laggards. Depending on the phase of adoption, different kinds of management focus are required. Crossing the chasm from the innovators to the early adopters, the early majority is often referred to as the Valley of Death. This is where 'management of adoption' is especially needed.

The process of how new products and services are adopted as an interaction between existing users and potential users is well described by the Bass Diffusion equation (Bass, 1969), and the diffusion equation is influenced heavily by two coefficients: the coefficient of innovation and the coefficient of imitation. The coefficient of innovation is a measure of the influence of the novelty of the product and the advertising impact while the coefficient of imitation is measure of the word of mouth effect where existing users refer the product or service to potential users. Typically, the coefficient of imitation is much more important for adoption success than the coefficient of innovation and the connectiveness that modern social networks provides further drops the barrier for referrals. Paying attention to the user experience of using an innovation can be really crucial in ensuring a thrilled user can influence further adoption by spreading the word.

# Chapter 10
# Agile Development and Production

Agile development and production is hugely important as development and adoption cycles shrink. The clock-speed of virtually every industry is being shortened by digital disruption. The objective of agile development and production is to move iteratively as quickly as possible from idea to a product or service, which meets a need or opportunity. Agile production is about creating and supporting the processes, training and tools quickly and iteratively create new products and services as well as responding to customer needs and market shifts. A key enabling factor is the development of a production support tools to allow the relevant organization members from designers, developers, and marketers to a common database of code and other parts. As different products cycle through the production process from concept to product or service, they draw upon and also add to the manufacturing support platform.

## 10.1 Agile Methods

There are a number of core principles for agile software development whereby requirements and solutions evolve through the disciplined collaborative efforts of self-organizing cross-functional team—these include proactive and adaptive planning, rapid evolutionary and iterative development, early delivery of system artefacts, continuous improvement all encapsulated in a closed loop, and high velocity repeatable process (Fig. 10.1).

The famous manifesto for agile development is a set of principles not published from the high moral ground of academia but from the swamp of practice and are evolved from real-world practitioners experience of what actually works. In the inherent tensions that are present in product development, the agile manifesto advocates biasing towards

- Working software over comprehensive documents
- Individuals and interaction over processes and tools

© Springer International Publishing Switzerland 2018
M. Curley, B. Salmelin, *Open Innovation 2.0*, Innovation, Technology, and Knowledge Management, DOI 10.1007/978-3-319-62878-3_10

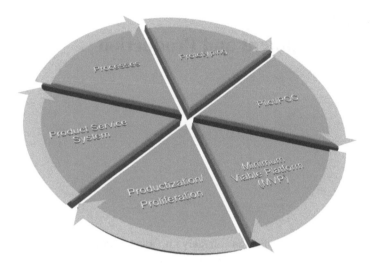

**Fig. 10.1**  Agile Production

- Customer collaboration over contract negotiation
- Responsiveness to change over strictly following a plan

Perhaps the core principle underpinning agile developed is that the highest priority is about providing high customer satisfaction through early and continuous delivery of value adding software. In this closed loop, iterative process, changing requirements are actually welcomed even late in the process as it means that the product or service developed will be closer to meeting market and customer needs. Using cross-disciplinary teams where developers and business people work together on a daily basis with software delivered frequently in periods from a couple of weeks to several months are also core to success. The secret ingredient in successful agile development is often passion, where individuals motivate and propel the team forward. Daily stand-up meetings and continuous face-to-face interaction are the modus operandi of agile teams and actual working software is the primary measure of success. Regularly creating the space to enable self-organizing teams to reflect on their processes and effectiveness is also important. Although 'Sprints' are the normal work method for agile teams, they need to be done in the context of a 'marathon' mindset in that sustainable competitive advantage is only achieved by consistently delivering a stream of successful software innovations.

Core OI2 process in Agile production include rapid **prototyping** and **piloting,** and these patterns are often used in the context of a living lab whereby rapid experimentation with users and citizens is a key part of the innovation process. The Intel Collaborative Research Institute for Sustainable cities uses London as its living lab and research projects/ experiments have been run in locations such as Hyde Park, Enfield, Elephant and Castle, Tower Bridge and across a number of different schools. Using an instantiation of the Intel ® IOT platform over six hundred sensor end-points were deployed. One disruptive technology opportunity where Intel partnered with Enfield council and several UK suppliers was to build an IOT-enabled distributed air quality system which could give the borough a much better picture of the quality of air in the borough and provide opportunities to drive actions to improve quality.

## 10.2   Design Science Research and Digital Assembly Lines

Design science research is an outcome-based digital research and innovation methodology, which is at the core of the OI2 paradigm. It is a six-step process, which offers specific process steps and has guidelines for evaluation and interaction within research and development projects.

The six steps are

- Visioning
- Understanding
- Implementing
- Evaluating
- Exploiting
- Results

These are six steps, which are connected in a closed loop fashion with frequent iteration as shown in the following figure (Fig. 10.2).

We begin with the vision steps where a problem or opportunity is to be approached. We start with clearly defining the research question or opportunity. It is critical to have a clear problem or opportunity statement to begin with and getting a multidisciplinary set of stakeholders involved. Creating a shared purpose or vision is crucial to success.

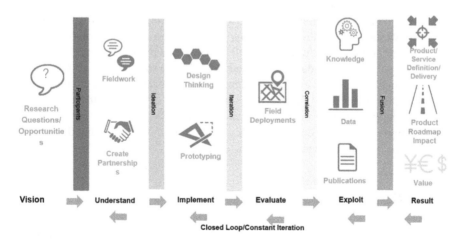

**Fig. 10.2** Design Science research

The next step is the understanding phase where both field and desk research are carried out and the right team is established to design and produce the innovation. Agile methodologies underpin the OI2 DSR approach augmented by techniques such as DeBono's six hats or lateral thinking and focused brainstorming.

The next step is implementing where agile methods create prototypes, which ultimately become a minimum viable platform, which can be both co-created and tested with lead users and early adopters. In the evaluation process, field deployments in living labs can be very useful so innovators and users themselves can see how well the solution meets the need or opportunity. There can be a range of outcomes from the evaluation, not limited to data, knowledge, patents, and publications. Through the ongoing iterative process, prototypes and MVP are constantly improved unless a data-based decision is made to stop the innovation, which is also a very valid outcome. The final step is to convert the innovation into value whereby a new product or service is defined and delivered or the outcome becomes part of a product or service roadmap. As we have said before, innovation only happens when there are adopters and value has been created.

In the iterative flow, we continually passed through phases of ideation, iteration, correlation, and ultimately where all the learnings are fused into a new product or service or indeed a valuable insight.

When such a process is well practiced and is supported by foundational tools, it becomes the equivalent of a digital design and assembly line.

## 10.3  Prototyping

Prototyping and experimenting are the core learning process in agile development and production and is a rudimentary or rough working model of a product or service, primarily built for demonstration and learning outcomes. A picture is worth a thousand words and a working prototype is worth a hundred pages of product/service requirements. Prototyping emanated from the automotive industry where manufacturers would built small-scale prototypes and indeed concept cars to quickly test and showcase possible functions and features. While prototypes are often used to see if the software being developed meets the software specification, prototyping can and should be used to evolve and develop the software specification. It also helps with estimating the fuller software development efforts and the prototype servers as the key input to the productization process. When a prototype is sufficiently mature, then it may be deployed in a pilot or proof of concept, but this stage often requires a minimum viable platform.

## 10.4  Minimum Viable Product

A working minimum viable product (MVP) is one of the holy grails of agile development teams, delivering a product or service, which has just the core features and functionality so a product can be deployed to a subset of early adopter possible customers who can give feedback and help, validate, or invalidate the product vision. Feedback on the MVP becomes the core stage gate on deciding to productize, otherwise modify, or possibly stop product development. A core objective of use of an MVP is to help pick winning products early, while avoiding building products that customer do not want or will not pay for and maximize efficient learning. The further a product or service goes through the development cycle the costlier it becomes to fix mistakes and feature or functionality mismatch. An MVP is the core artefact in the process of product/service idea generation, prototyping, evaluation, and ultimate productization.

**Case Study: IOT MVP at Intel Labs Europe**

While I was at Intel Labs Europe, we effectively used a design science research and agile development process to advance a concept and vision around an Internet of Things (IOT) platform. In collaboration with Charlie Sheridan who was my head of IOT systems labs, we established collaborations with both the cities of London and Dublin. We led the establishment of the Intel Collaborative Research Institute in London, based on OI2 principles with two of the leading universities in the world, Imperial College and University College London to create two compelling and credible living labs. Working also with leaders from the cities we established a context of high trust and were able to deploy and test multiple IOT uses, cases based on a common MVP IOT platform.

The figure describes an 8-week process we used at Intel Labs Europe to go from a concept to a working IOT Gateway MVP which we deployed in Dublin City after 6 weeks and then proliferated to a London City living lab. The process began by hosting representatives from several Intel divisions including the IOT and datacentre product group at our lab in Leixlip. In the first week of the agile process a shared vision and architecture of what we would jointly produce was developed. The goal was to develop an IOT platform, based on Intel assets and new components which we would need to develop, that could be deployed in cities and sense a wide range of measurements from microclimate to air quality. Based on the shared vision and rudimentary architecture, an integrated team of software developers, hardware architects, and business people led by our lead principal investigator/scrum master Mark Kelly moved quickly through a series of high intensity sprints, an example of which is shown in (Fig. 10.3). This culminated in a working prototype being deployed to the living lab we had deployed in Dublin. Following 2 weeks testing and iteration, we could declare that we had an MVP and then deployed in wider scale in the living lab that we had established in London in collaboration with UCL and Imperial College London. The agile process proved to be highly effective and the MVP and learnings from it provided the main input to the Intel ® IOT platform architecture.

| WR & DCSG Big Data Team on site | Sprint 1: Gateway Development<br><br>*Quark + IDP2.0* | Sprint 2: Cloud and Device Messaging<br><br>*Arduino on IDP* | Sprint 3: System Management +Interface Development | Sprint 4: Dublin Live | Sprint 5: London Live |

**Fig. 10.3** Agile process for Intel (r) IOT Platform MVP

## 10.5   Living Labs

In OI2, we see a new form of industrial research where the use of living labs where real-time experimentation is conducted in real-world situations allowing simultaneous technical and societal innovation. Intel Labs Europe living labs in Dublin, London and San Jose are good exemplars of OI2 in practice where city government, industry, academia, and citizens are involved in the innovation process (Fig. 10.4).

In the London Living Labs using a common Internet of Things platform, multiple innovative projects were deployed with engagement across many stakeholders and we are witnessing embryonic network effects as the platform gets adopted in new unanticipated scenarios.

The probability of breakthrough improvements increases as a function of diverse multidisciplinary experimentation, which is an essence of OI2. In today's complex world, experiments simply cannot be conducted in isolation. Collaborative research will accelerate the innovative process and improve the quality of its outcomes. While closed-world innovation will not disappear, it will be dwarfed by the efforts of cross-border and cross-disciplinary teams that enable a wide spectrum of stakeholders to take on active roles.

### 10.5.1   Living Labs in a European Context

The origin of 'Living Labs thinking' was in the Massachusetts Institute of Technology (MIT), where the approach was to construct test and verification environments in laboratory settings to develop and experiment with different technology solutions with real users invited to visit those environments. However, some earlier attempts exist to enhance users and other stakeholders for developing, validating, and testing products and services in living labs and living laboratories in the USA. MIT approach led to early prototyping with 'real users', again with the potential for faster scaling-up of the results. William J. Mitchell, a Dean and Professor at MIT, was one of the key drivers in this new research and prototyping approaches.

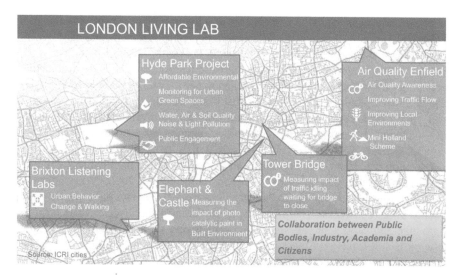

**Fig. 10.4**   London Living Labs

When discussing that approach from a European innovation-system perspective, it soon became evident that end-user involvement could be *the* key factor for the renewal of the European innovation system. In Europe, we have very demanding but also very diverse user communities for our products and services; the question was: how do we harness these user communities to increase the success rate and speed of the innovation processes in Europe? Part of the answer came from *Democratizing Innovation* by von Hippel (2005), which triggered thinking of co-creation and user involvement in the innovation processes.

In 2003, an industrially led think tank for the living labs strategy in Europe had been established in liaison with the European Commission's Directorate-General for the Information Society to develop and conceptualize the European approach. Soon it became evident that the European approach should be focusing on creation of innovation hubs, which would build on the quadruple helix innovation model: strong and seamless interaction of industry, the public sector, research institutions and universities, and last, but definitely not least, 'the people'.

The goal was to create an environment that would attract industrial and research investment due to better innovation dynamics than could be achieved with traditional approaches. These dynamics would be supported by the public sector, and one of the focus areas would be public sector services that could be co-developed with the user communities in real-world settings. Part of this thinking was based on the idea of stretching the boundaries of societal behaviour, given that we saw, the connectivity and environments with shared information communications technology (including emerging social media) begin to change society. The quest was to push the boundaries with real-world projects featuring strong technological development. Only by doing the research and development with citizens, could we see what solutions would ultimately be acceptable and thus scalable to products and services.

These early steps led to the first concept of *a* living Lab in a European context: a real-world site, not an extension of a laboratory, with sufficient real-world users numbering at least in the hundreds to allow for scalability. By this point, we know that a European living Labs should comprise citizens, application environments, technology infrastructure, organizations, and experts. Later, we would add societal capital into the picture once we learned that living labs depend on the idea of spill-over effects back to society. This spillover provided motivation for all stakeholders—including citizens—to contribute to the common goal, making living labs a winning game.

Based on these concepts, the European Commission and the Finnish EU presidency launched the first wave of European living labs in 2006. Next came the European Network of Living Labs (http://www.openlivinglabs.eu/), which grew quickly under subsequent EU presidencies to the substantial scale it has now, to more than 340 sites, some of which are even beyond European borders. And, the network is still growing. Living labs now have a strong foothold in all European regions, and the approach is also being applied as an important component of regional innovation systems.

On European level, the networking associated with living labs is of utmost importance. Using a living labs methodology to find common, scalable solutions with different user environments is essential when driving towards common European services based on common architectural approaches. I am happy to see that the thematic cross-border networking of the sites is accelerating, enabling the most interesting living labs to collaborate as partners, for example, in Horizon 2020 projects, especially in the contexts of smart cities and public services.

### 10.5.2   *Open Innovation as Part of Living Labs Thinking*

Openness was a starting point when developing living labs in the European way. Openness in sharing platforms for services but also an open mindset for collaboration amongst all stakeholders. The thinking stems from the early 1990s, when the hot topic was 'virtual' and 'holonic enterprises' were creating both agile and scalable structures for operations, by sharing common operating architectures and by collaborating strongly on a task-driven basis. Good examples of holonic-/fractal-/virtual-enterprise theory were developed, for example, in the intelligent manufacturing systems initiative amongst the leading industrial economies. By scaling up this thinking, we come very close to the foundations of Living Labs by adding in the public and societal components.

Combining the user-driven and co-creation approaches to innovation from von Hippel with the open innovation approach Chesbrough (2003) introduced, we come to the two fundamental aspects of modern innovation theory. The definition of open innovation by von Hippel focuses on the creation of public goods, whereas the one by Chesbrough builds on sharing, cross-licensing, and in that way is a market- and product-driven approach.

Open platforms, sharing, and seamless interaction between all stakeholders are essential characteristics of living labs. The quadruple helix has thus been central as an innovation model underlying the Living Labs movement from the very beginning. And, recently, we are seeing a new paradigm as open innovation ecosystems are increasingly becoming the synthesis of Living Labs and open innovation processes. Open innovation has become much more than the cross-fertilization of ideas between organizations; it has become a flow of colliding ideas, raising sparks for new innovations in real-world settings.

## 10.6  Privacy by Design

Privacy by design (PbD) is an approach to agile development which takes privacy into account during the whole development cycle, but particularly at the design cycle so that privacy is not an afterthought and is built into the innovation. PbD asserts that privacy assurance must become part of an organizations' standard modus operandi rather than being about compliance with legislation and regulatory frameworks. PbD was originated by Dr. Ann Cavoukian who was the then Information and Privacy Commissioner in the 1990s and it is less about data protection but more about designing so data does not need protection. A reference example of this is Dynamic Host Configuration Protocol where computing devices based on random identifiers gets an IP address from the server and then is enabled to communicate without sharing personal identifiers. The worldwide GPS system is also quoted as an example where a device can detect its geographic location without sharing identity or location.

Many applications will use user data, natively, aggregated or anonymized and increasingly regulators will look for privacy impact assessments before such services as developed and launched. PbD advocates for an approach that is proactive rather than reactive. Rather than waiting for privacy risks to occur, PbD aims to prevent them from occurring. Ann Cavoukian has articulated seven principles of PbD which have been widely adopted which can be summarized as follows:

1. Proactive and preventative rather than reactive and remedial
2. Privacy as the standard and default setting
3. Privacy embedded into design
4. Full functionality—positive-sum, not zero-sum
5. End-to-end security—full life cycle protection
6. Openness—visibility and transparency
7. User-centricity, respect for user privacy

Organizations or developers who ignore or skirt around privacy will ultimately be found out. The coming into law of the 2018 EU General Data Protection Regulation will be a regulatory milestone that should change the face of privacy in the future with likely other jurisdictions following suit quickly.

## 10.7   Productization and Proliferation

Perhaps the most difficult step in the agile innovation cycle is the productization process, where disciplined and robust processes for launch and support of a product or service are necessary. This means taking a manufacturing and quality assurance mindset and having the appropriate processes, capacity and scale in place to drive adoption and meet demand. As we have discussed before, it is crucial that the innovation has been designed with adoption in mind. As more and more products and services are cloud enabled, the ability to have cloud, bursting capabilities to meet unanticipated demand is hugely important. With fickle consumers, a service only has to be unavailable once and it can then be dismissed forever. If a product or service has been 'designed with adoption' in mind, then the process of proliferation is made much easier. With the marginal costs of distributing software almost zero, digital products are designed for low cost distribution and particularly those that use Apps stores have dramatically reduced adoption barriers. Social media can be a hugely important tool in the proliferation of digital solutions.

### 10.7.1   DevOPs

DevOPs is emerging as the new way of managing seamless transition and operation between development teams and IT operations, which need to support and monitor production and software services. Less than a decade old, the term DevOps emerging from the agile community in Belgium in 2009. DevOPs sits at the intersection of software development, IT operations, and quality assurance.

DevOps is about establishing a culture, infrastructure, and environment when software-based products can be built, tested, released, and adopted seamlessly and with high velocity. The benefits of a DevOPs approach are faster time to market but in particular lower failure rate of new releases. With increasing complexity of software and increased mission criticality of software strong discipline and processes, need to be in place. An over the air download to the embedded controller of a moving vehicle cannot fail.

## 10.8   Servitization

An observable shift from 'product' to 'service' is happening. Aston business school define servitization as 'The process by which a manufacturer provides a holistic solution to the customer, helping the customer to improve its competitiveness, rather than just engaging in a single transaction through the sale of a physical product'.

Many companies seek to build on services to products and this while also can be financially attractive, will also help with the establishment of a new sustainability paradigm: A simple example of this is the adoption of cloud computing which enables new service models which are likely more efficient and effective than each company provisioning their own hardware and infrastructure. Additionally, cloud computing will also help shorten development time for new services, helping to bring benefits faster to the market and to the broader society. Servitization is about moving from selling products to services and indeed even today information about products is sometimes even more valuable than the products themselves. Companies such as Rolls-Royce with their 'power by the hour' strategy have pioneered this approach. Instead of selling aircraft engines, Rolls-Royce sell hours of flight time in a win-win configuration for Rolls-Royce and their customers. Similarly, Amazon's and Apple's transformation of the book and music industry are other great examples of servitization being profitable, more convenient, and more sustainable.

Servitization can be as the ultimate for extended products, which consist of tangible and intangible components. In this approach, one can see the tangible part giving access to the intangible part of the extended product. Good examples of this can be found in the telecom market where in certain markets the phone (device) is paid by rather high subscription fees, or in the reverse case in the Apple iProduct approach where the premium price of the tangible component is giving access to (relatively) cheap contents.

Extended products are often a successful path towards servitization. Servitization fits beautifully with the sustainability paradigm, which is about moving from maximizing consumption to optimizing asset utilization and asset longevity. Some automotive manufacturers are exploring models, which exactly embrace this concept. Servitization is also attractive from a business model standpoint with the opportunity to move from one time payment to annuities. Still the life cycle cost from the user perspective needs to be looked at more thoroughly, case by case.

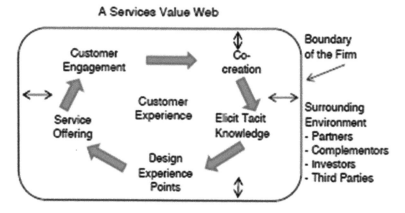

**Fig. 10.5**  Services Value Cycle, source Chesbrough

Good examples of servitization are, e.g. the concepts leading to Mobility as a Service: MaaS. The service providers integrate the offerings under a common umbrella and the user subscribes to a mobility package. Conceptually, and first pilots are ongoing in the Helsinki capital region, where the detailed planning is going on on a subscription-based mobility scheme. The user can subscribe on monthly basis on packages for different transport means (bus trips, train trips, taxi, bike rental, and collective bus services) depending on the need. Ministry of Communications in Finland is in active phase to launch the service in 2016. This is also an example where traditional business models of the individual service providers are challenged as the concept is based on collaboration rather than competition.

Henry Chesbrough has sketched an elemental services value chain or indeed web, which is based on design experience points, which leverage co-creation as, described in the figure below (Fig. 10.5).

# Chapter 11
# Industrial Innovation

*'No matter how beautiful the strategy, you occasionally need to look at results'.*

—*Winston Churchill*

Too often innovations fail. Industrializing innovation is about managing innovation as a capability similarly to manufacturing processes, with the outcomes being a series of more predictable innovations (Fig. 11.1). By applying an industrial lens and discipline to the Innovation process, the probability and predictability of innovation outputs and outcomes can be significantly increased. There is of course a tension between the risk minimization needed for reliable manufacturing and the risk tolerance needed to encourage innovation; however, these can be appropriately balanced to deliver more predictable outcomes.

In this chapter, we will review key attributes of core functions that are part of the industrial innovation pattern where we begin with visioning.

## 11.1 Visioning

A core function of an R&D department or product development department is producing a vision, which provides a context in which team-based innovation can take place. A strategic innovation vision can provide a lens for driving both more effective and efficient innovation. In OI2 terms taking the users as co-creators help market adaptation and fast scale-up of products and services to be developed. This safety net not only speeds up the process but also makes the likelihood of market success higher.

## 11.2 Strategic Innovation

Innovations are likely to be one off or sporadic unless there is a vision, which paints a desirable future state. The most effective kind of innovation happens with in the context of a vision, and this creates the possibility of a stream of linked and mutually reinforcing innovations. Strategic innovation happens when we have ideas,

© Springer International Publishing Switzerland 2018
M. Curley, B. Salmelin, *Open Innovation 2.0*, Innovation, Technology, and Knowledge Management, DOI 10.1007/978-3-319-62878-3_11

**Fig. 11.1** Industrial innovation

which are implemented and then adopted, and all of this takes place in the context of a vision. We could loosely define strategic innovation as the product of ideas by execution by adoption, all performed in the context of a vision.

$$\text{Strategic Innovation} = \text{Vision}\left(\text{Ideas} \times \text{Execution} \times \text{Adoption}\right)$$

Europe 2020 is an example of a pan-national vision, which creates a context for vectored innovation; however, there is recognition that its predecessor the Lisbon Strategy while a compelling strategy failed to have the execution and adoption to make it real. The development of USB is a great example of a strategic vision that had massive strategic impact.

**Case Study: USB Invention to Innovation**
The development of the Universal Serial Business (USB) standard by Intel and others is an example of OI2 at work although at the time of the USB initiative the concept of OI2 had not been conceptualized or described. Prior to the creation of USB computers required many different drivers and ports with different interfaces to communicate with peripheral devices. Ajay Bhatt at Intel created a vision, which became a shared vision of a new standard connection interface that would recognize and immediately be able to use new devices that were plugged into the PC. This shared vision became a reality, which dramatically drove the development of many new peripherals, and dramatically improved the ways files could be shared from user to user through USB sticks. Collaborative friction was dramatically reduced in that users could immediately plug-in devices such as web cams, smart phones, even

scientific instruments to PCs, quickly, seamlesslys and instantly. More than ten billion USB devices have been shipped making it one of the most important computer interfaces ever developed. The USB protocol acts as a translator between various devices translating bespoke instructions into a standard set of instructions that a PC can understand.

Even though a vision can be compelling sometimes, adoption is slow, in this case because USB required a redesign of the Microsoft Windows software platform that was not available until 1998. In the meantime, Apple computers launched its iMac, the first mass-produced USB-enabled computer that helped catalyse adoption and acted as a proof point amongst a nice user community. Subsequently, an explosion of innovation happened with thousands of USB products created creating all sorts of innovations. Because a new platform with well-defined standards was created, a completely new ecosystem grew up around developing USB based products, and the user market for PCs and iMACs also grew very substantially. The USB became a key to unlock massive latent potential of the PC.

Intel Corporation was granted a patent on USB technology in 1997, but the protocol system was created in conjunction with six other companies and the technology rapidly became a standard in the industry. However, we know from both legal, business and technical perspective that patents and industry standards can be tricky. The solution used for USB was to create a 'patent pool' where all the holders of essential USB technology patents combined their patents into a single pool, the USB Implementers Forum. This open mechanism allows other organizations who wished to use USB to license USB patents under reasonable and non-discriminatory (RAND) terms without having to pay royalties and without fear of litigation from the patent holders.

This OI2 spawned a massive USB development ecosystem and created value and contribution far beyond the value of the USB ecosystem itself.

## 11.3   InVenting

Since the dawn of time, people often have been inventing and we all have heard the expression necessity is the mother of invention. Inventing is about something new into being. Many inventions are useful but equally many are not. We often confuse invention and innovation, many inventions are not successfully adopted, and it is often only through the perseverance of an inventor that an invention is ultimately adopted. Innovation is a much broader process thinking about the broader aspects of getting an invention to market including the business model, marketing plan, and production process for the invention.

## 11.4  Validating

Validating is a process where we test to confirm or disprove assumptions around the particular functionality and features of a particular invention or innovation. Today, a key mantra around validation is 'Fail fast, learn fast'. By experimenting quickly, we can learn which research project or innovations we should stop or shut down, as they do not meet the requirements. This is a hugely valuable process as it can avoid a company investing in a product and only finding out that it does not or will not work when it appears on the market. Validating also allows us determine which innovations are likely to be successful so they can be accelerated towards the market.

## 11.5  Venturing

Venturing is a great way of trying to accelerate and harness the velocity of an ecosystem, and it can be a two-way process. Many companies have initiated corporate venturing program to get competitive intelligence and make investments in companies, which have the potential to strengthen their ecosystem and/or may yield disruptive technologies or business model. In parallel, some companies have put processes in place to support the creation of ventures from within an R&D department or indeed from any source in the company. At Intel, a New Business Development process, organization, and fund were put in place to support this source of innovation where promising ideas or technologies developed internally can be accelerated trying to commercialize. Often a 'Learn fast, fail fast' mantra permeates these kind of models. Notable successes in Intel include Apple Thunderbolt, which went direct from Intel Labs to Apple—Thunderbolt was a revolutionary I/O technology that supports high performance data devices and high-resolution displays through a compact port. A different type of success was the pioneering work of former Intel Fellow Mario Paniccia on Silicon Photonics for high-speed datacentre interconnects, leading to the development of silicon-based semiconductor lasers to support 50 GBit/s optical links for servers. His work graduated from the silicon photonics lab through a venturing process through to becoming a silicon photonics solutions group in Intel's highly successful datacentre business group.

From a shared value standpoint, there are lots of attractive things from a start-up's point of view to being involved in a corporate venturing process. These include five C's Cash, Cache, Capacity, Capability, and Confidence. In early stage ventures, cash investment is of vital importance but other factors can be of equal if not more importance. The 'cache' of having an investment from a leading company is important and can inspire 'confidence' in others to purchase products or services from the start-up. Equally, the start-up can leverage capability and capacity from the venturing company.

**New Forms of Venturing Case Studies**

OI2 formations in venturing are emerging with the example of Cisco, Intel, and Deutsche Telecom's 'Challenge Up' accelerator program being particular noteworthy. Challenge up is a start-up and accelerator program for early stage IOT companies. Typically, the process is run yearly and approximately 12 companies are chosen from the hundreds that apply and are run through an intense 6-month incubation program with physical meetings typically held monthly in different European start-up hot spots. The three companies align around a shared purpose of driving open innovation in the Internet of Things. Start-ups benefit from the technology, business, and sales expertise and reach of the three companies resulting in accelerated development and faster go-to-market.

Mastercard's Start Path program enables start-ups to leverage Mastercard's global ecosystem and to leverage Mastercard's capability and expertise, creating new relationships with Mastercard and some of its key customers and indeed suppliers. Mastercard involves key customers and partners in the process through a subscription model, which helps fund the program. Companies, which are selected for Start Path, participate in a 6-month virtual program with 2 immersion weeks in different cities and importantly there is no upfront equity required for participation. Companies can apply to regular selection competitions and typically receive much benefit from participation in the program including the ability to network with other similar stage companies and Mastercard executive and technologists.

## 11.6 Velocity

Today, the clock-speed of most industries has been dramatically speeded up as information technology more and more permeates new products and services and the R&D and production processes used to create those products and services. Time-to-market is crucially important and one takes a risk by waiting until one has a perfect product. The risk is that in a competitive market, a competitor may usurp your organization by delivering an inferior product earlier into the market and achieves a strong market share or even becomes a dominant design by virtue of being earlier in the market. Velocity is of course more than random speed, speed being in a particular direction. When an organization has velocity in the context of a strategic innovation vision, then there is a much higher probability of sustained success.

## 11.7 Value

Leading R&D organizations are increasingly thinking about value earlier and earlier in the innovation process. IT organizations, which have often assumed central roles in the digital transformations of many organizations, now often focus and design their digital innovations based on value dials. While I was at Intel, we developed a practice of measurement around the so-called value dials. At Intel, value dials represented a standard set of business variable for which innovation efforts could be targeted against to drive improved business performance, for example, days of inventory or factory uptime. By maintaining a list of business variables and the current monetary value of improving performance of these, innovations competing for funding could be compared and also managed to success to deliver on the benefits, which were promised at the start of the innovation project.

## 11.8   New Innovation Value Constellations

When looking at new collaborative forms between stakeholders and enterprises, we see a dramatic change from predetermined collaboration to more opportunistic approaches based on mutual strengths. The figure below illustrates this (Fig. 11.2).

Typically, subcontracting chains are stable, within clusters of industries. The component supply chain is stable, and very much determined top-down, also regarding the pricing and the design of the components. The subcontractors provide mostly the production capacity.

In many areas, we see the subcontractors becoming stronger and having responsibility of managing larger entities. Typically this can be seen, e.g. in automotive industry where suppliers are serving many brands with their products and services. The subcontractor typically is offering special competence and capacity to be delivered. In this context, the clustering includes strong networking effects. However, the business relations are more or less fixed and predetermined.

When the business opportunity dynamics is even higher, and the production flexibility increases we move towards business model constellations, where enterprises/competencies configure according to the needs the task to be performed is setting.

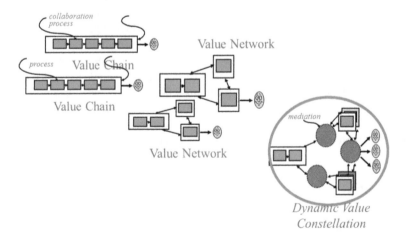

**Fig. 11.2**  Towards Value Constellations

This value constellation analogy should be seen as stellar constellations: The constellation is dependent on time, the viewer, and the location. We are moving to task-driven forms of delivery, which is very much in the spirit of Open Innovation 2.0 as well, where dynamic resource and competence configuration is determining the success of a company/ecosystem.

## 11.9 Industrializing Innovation: Innovation Capability Management

A key component of the Open Innovation 2.0 paradigm is explicit management of innovation capability and systematic management of the improvement of the capability. Organizations are increasingly realizing that innovation is not something that happens always by accident or serendipity.

## 11.10 Innovation Systems

In organizations, we can design closed loop innovation systems and look to manage innovation as a capability. Building on an innovation system, we can articulate an innovation capability maturity framework as depicted in the following diagram for the four vectors involved in the innovation management system.

As one traverses the innovation capability maturity framework, one moves through phases of practices that move from ad hoc behaviour to current to best to next and finally future practice. By systematically managing capability organizations and ecosystems can improve the predictability, probability, and profitability of innovation outcomes.

The Innovation Value Institute has built a sophisticated innovation assessment instrument, which allows companies to assess their relative innovation maturity and provides recommendations from a best practice repository, which contains leading-edge practices from companies and academics. The build out of the innovation CMF in itself an example of Open Innovation 2.0 at work with more than eighty organizations working together to create shared value. The figure below shows the innovation capability building blocks that the IVI has identified and that need to be in place to manage innovation more systemically (Fig. 11.3).

### 11.10.1 Innovation Strategy and Innovation Capacity

Innovation strategy and innovation capacity have to be aligned with business strategy as the following figure shows (Fig. 11.4). If innovation capacity is not aligned with innovation strategy and it itself is not aligned with business or organization strategy then you get innovation leakage and wasted efforts. In an organization, it is

| Innovation | Managing Innovation | Funding Innovation | Managing Innovation Capability | Managing and Assessing Innovation Value |
|---|---|---|---|---|
| 5. System Innovation | Continuous realignment of pipeline | Amplified budget | Innovation excellence | Predictable, Probable, and Profitable |
| 4. Managed Innovation | Explicit strategy | Co-funding with the business | Infrastructure integrated | Proactive change management |
| 3. Defined Innovation | Management commitment | Formal budget allocation | Infrastructure established | Active change management |
| 2. Sporadic Innovation | Tactic tolerance | Project-based allocation | Occasional skunk works | Informal assessment |
| 1. Initial/Ad-hoc Innovation | Ad-Hoc | | | |

**Fig. 11.3**  Innovation Capability Maturity Framework (CMF)—source Baldwin and Curley

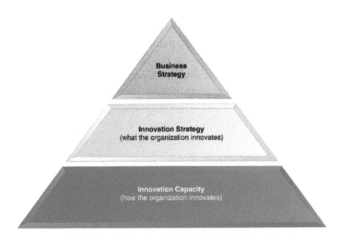

**Fig. 11.4**  Innovation Capacity, Strategy, and Business Alignment

important that there is appropriate governance so that as innovation capacity is developed, it is applied to the organizations innovation strategy, which in turn needs to be aligned with the overall organization strategic direction.

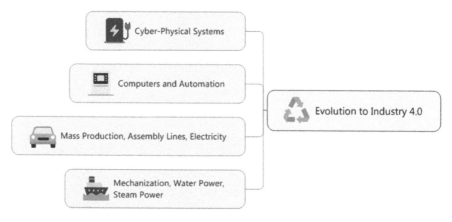

**Fig. 11.5**  Industry 4.0

## 11.11   Industry 4.0

Industrial innovation has been at the heart of driving progress since the dawning of the industrial revolution, and now we are witnessing what many call the fourth industrial revolution or Industrie 4.0 as originated by the German government and promoted by the German Academy of Engineering Acatech (Fig. 11.5).

The first industrial revolution began towards the end of the eighteenth century with mechanization water power and steam power being the key drivers. At the start of the twentieth century, we entered the second industrial revolution with the emergence of mass production, as exemplified by the Model T Ford and assembly lines and electricity as a new power source. The third industrial revolution began in at the start of the nineteenth century with the adoption of computers and automation. Arguably, the fourth industrial revolution has begun with the simultaneous emergence of the Internet of Things and ubiquitous networking. The use of cyber-physical systems enables ever more complex and capable manufacturing to occur.

# Chapter 12
# Data-Driven Innovation

Big data and data-driven innovation are creating significant information monetization opportunities. Former European Research Commissioner, Maire Geogeghan Quinn, coined the phrase 'Knowledge is the crude oil of the 21st century' and it aptly describes the opportunity. Even this analogy is not powerful enough as data or knowledge does not get used up as it is shared or used up, as it is a non-rival good. In fact, it often multiplies. Debra Amidon, Piero Formica, and others have defined the three laws of knowledge dynamics, which underpin the practice of data-driven innovation.

For many individuals, organizations, and ecosystems the emergence of big data, with ever more powerful machines and data mining and machine learning algorithms will present an opportunity of a lifetime. However, the opportunity of a lifetime has to be taken in the lifetime of the opportunity.

Data-driven innovation is simply about using data for innovation. Here, we identify six key patterns although we recognize that there are other patterns and more will be discovered. Data is a non-rival good meaning that it does not get used up or dissipate unlike a rival good such as petrol in a car. Data has a very unusual property in that the same data can be used in theory in infinite ways and by an infinite number of users to generate insights, services, and value. Thus, data can be considered as both an infrastructure and a capital good.

Knowledgeification in fact changes the economy, as all our economic rules have been based on managing and optimizing scarce resources, now we are entering an era where there is an abundance of the primary wealth generating resource, data. Indeed, the so-called GAPP rules seem no longer appropriate in that for companies such as Google and Facebook their physical assets on their balance sheets only account for approximately 15% of their market capitalization or worth (Fig. 12.1).

When looking at fast-growing companies, the so-called unicorns, it is worthwhile to observe that they are based on user-generated contents and large user communities, signifying their role as data generators for the new value creation business models.

© Springer International Publishing Switzerland 2018
M. Curley, B. Salmelin, *Open Innovation 2.0*, Innovation, Technology,
and Knowledge Management, DOI 10.1007/978-3-319-62878-3_12

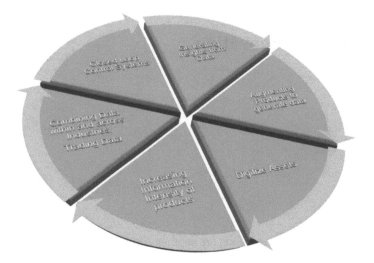

**Fig. 12.1**  Data-driven innovation patterns

Peter Pham (2015) describes the need for a new paradigm well '*A shift in thinking needs to occur in the way that data is viewed. Data is no longer a static disposable resource that loses usefulness once it has served its singular purpose. Its life may be extended through multi-use, multi-purpose data processing. As a renewable resource, its value should be assessed not by the bottom line, but as an asset that not only grows in value but one which further provides value creation opportunities. It is the raw material of business and as with other raw materials i's ability to be used for a variety of applications makes it profoundly more valuable than the original product itself*'.

## 12.1  Generating Insights from Data

A great example of generating insights from data is Farecast, which was an early consumer applications of big data whereby using pattern recognition Farecast would inform consumers to pick the likeliest time to purchase a particular flight between two destinations or could be informed when a particular flight was about to increase or decrease in price. Microsoft paid $114 million for Farecast in November 2013, which gave the two founders a handsome return on their investments. Farecast is an exemplar of the idiom 'Overnight success takes on average a decade' as it was founded in 2005 before such kind of services became mainstream and adopted by other travel sites such as Kayak. It demonstrates the importance of timely pathfinding and technology scouting. Machine learning and artificial intelligence are emerging as related disruptive technologies but are being demystified by technologists such as Irving Berger describing these as simply 'brute force engineering' whereby big data, ever most sophisticated software and ever more powerful computers allow better and better results. As Arthur C. Clarke has said 'Any sufficiently developed technology is indistinguishable from magic'.

## 12.2 Augmenting Products/Services Using Data from Objects

The second pattern is using data the objects generate, or indeed could generate to create new services and value. A textbook example is that of Rolls-Royce's power by the hour whereby using advanced sensing and analytics Rolls-Royce were able to transform its business model and sell hours of flying time as a service to airlines rather than just selling aircraft engines. With reduced cost sensing, volumes of new data and advanced telematics, Rolls-Royce were able to perform preventative and predictive maintenance of their engines, improving performance and reliability and extending engine life. This is an example of what is called servitization or product service systems where one is able to extend a product offering with a one-time sale to a service which has an annuity stream. Such an innovation is an early example of OI2 work as a new real win-win scenario was created for the supplier and the customers and the resulting solution was more sustainable than the previous offering. With the 'Power by the hour' service, cash strapped airlines were able to buy a service rather than having to carry significant capital investments on their balance sheets while Rolls-Royce better balanced and augmented their revenues. For the environment, the engines had better utilization and longevity and improved fuel efficiency as engineers could analyse big data from the engines to suggest modifications or future improved designs. Rolls-Royce was awarded a 2016 EU Innovation Luminary award for their servitization innovation. Data from supermarket loyalty cards are another example of generating value from consumer shopping data, supermarkets are better able to understand trends and different market demographics, tailoring inventory and tailoring offers to specific consumer based on the data available. The subsequent reduction in food wastage and overstocking is another example of a small step towards sustainable intelligent living. Renault use engine and driving data to coach drivers to drive in a more eco-friendly way, caring for the environment and also extending car and engine life.

## 12.3 Digitizing Assets

At Mobile World Congress in 2013 William Ford, CEO of the Ford Motor company said that the car is now part of the network. Others have followed on to say that, a car is now actually a datacentre on wheels.

With objects being addressable on the network whole new business models such as Uber and Airbnb have been created which are again examples of OI2 and are textbook examples of the sharing economy at work. While some regulators and incumbents may not like the new business models, one cannot argue that they are steps on the way to sustainable intelligent living whereby under-utilized assets are matched in real time to consumer needs in a way that creates a win-win situation for both parties in the value exchange.

Both Uber and Airbnb have shared vision and are examples of multi-sided markets, connecting producers and consumers efficiently and effectively. Also Airbnb is another reminder that most innovations emerge from a stated need, the two founders created the concept when they needed to generate income through letting a room in their apartment.

## 12.4  Increase Information Intensity

Marc Andreesen famously said that software is eating the world. Driven by silicon, software and network economics, and capabilities, it is now possible to dramatically increase the information intensity of products.

As everything from machines to cities become more instrumented and real-time communications continue to improve, these objects or entities become programmable and become software defined.

And as control loops consisting of acquisition, analytics, and actuation are increasingly easily constructed, performance of the machines or cities can be increasingly optimized. Indeed, in some cases information about a product can be more valuable than the product itself. A recent estimate showed that software content of cars was over 40% and Henry Ford said in 2017 that the future of the car was much more about software than the hardware.

## 12.5  Data Mining, Combining, and Refining

Data mining is the process of searching for and discovering patterns in large data sets involving methods at the intersection of computer science, machine learning, databases, artificial intelligence, and other domains.

Imagine the power of being able to combine, for example, Tesco consumer spending habits data and a national health service database of illness and using advanced analytics to look for patterns and insights. The opportunities are very exciting in generating recommendations for healthier buying habits as well as improving quality of life and saving on healthcare costs. This would be a great exemplar of OI2.

Intel Labs Europe (ILE) in collaboration with the UK Future Cities catapult and the greater London Authority, UCL, and Imperial College established a network of IOT gateways across different areas of London to collect various environmental data such as air quality and micro-climate information. Together with Enfield Borough, the air quality data was used to build a better picture of air quality across the borough to help inform real-time traffic management systems and future planning actions for the borough. ILE also built a prototype dynamic congestion charging system, which would change real-time tariffs for congestion charging as well as modulate incentive pricing for park and ride facilities based on the real-time air quality across a borough. When data is combined like this, there is the opportunity to create high precision, high frequency control loops, providing better services and taking steps towards achieving sustainable intelligent living.

## 12.6   Trading and Monetizing Data

Data monetization is the process of generating revenue by selling data available from real time of offline data sources. Financial services companies are good examples of firms trading or monetizing data where, for example, credit card issuing banks use customer transaction data to improve targeting of cross-sell propositions.

Another well-known example is whereby cellphone operators can sell their data so that it can be combined with navigational data to identify areas of traffic congestion and suggest alternative more efficient routes. Again, this type of example is an example of sustainable intelligent living whereby commuters get more efficient and faster commutes in a way that is more friendly to the environment.

## 12.7   Closed Loop Control

Our ever-increasing connectivity, the ever-increasing power of compute and the emergence of the Internet of Things where everything from cars to electrical substations to washing machines is creating the opportunity to introduce high frequency, high precision closed control systems to societal systems which were previously in Open Loop. For example, the electrical grid has been designed as a one-way linear system where energy is generated in bulk capacity and then distributed (quite inefficiently) through high voltage, medium voltage, and low voltage distribution systems. With the increasing availability of local renewable energy (wind, solar, etc.), smart home systems and smart heat storage systems, the opportunity exists to redesign the grid creating value for all participants, lowering costs and making the overall solution more sustainable. One Horizon 2020 project led by Glen Dimplex with a set of stakeholders from across the energy value chain from generators to consumers called real value is researching and demonstrating this across 1250 homes in Germany, Ireland, and Latvia. This model is an example of the emerging concept of collaborative consumption.

At the core of these kinds of innovations are the twin ideas of systems of systems and closed loop control through enabling functions of acquisition, analytics, and actuation. Data is acquired from a thing or system, and then analytics are performed to provide decision support, which then can drive actuation to change parameters to effect service improvements or efficiencies. The integration of these three capabilities enables the creation and operation of high frequency, high precision, management control circuits. An example of such a system of systems would be a dynamic congestion charging system in a city which dynamically updates congestion charging based on parameters such as localized air pollution, weather, and traffic measurements to help optimize real-time traffic flows improve commute times while minimizing environment impact. The operation of such a system will also create a lot of big data and the use of machine learning and offline analytics can create a second-order feedback loop, which can drive further system improvements based on insights garnered.

# Chapter 13
# Openness to Innovation and Innovation Culture

*'It is not just about open Innovation but openness to Innovation'.*

Open innovation requires courage to be open for innovation. It requires courage to seek (and discover) the new which might lead to disruptive solutions. Openness to innovation also has a twin: courage for seeking the unexpected. Experimentation involves failures but not fundamental and costly ones, and thus the probability of finding timely scale-up solutions is significantly higher than in traditional project approach.

The culture of an organization or indeed a society is a critical factor in innovation success. Culture can be hard to define but one of the better definitions is that of collective programming of the mind that differentiates or distinguishes an organization or society from another. Culture is complex and is a product of the cumulative knowledge, beliefs, experience, rules, etc. acquired and put in place by an organization or indeed a society. According to Dictionary.com, culture is 'The sum of attitudes, customs, and beliefs that distinguishes one group of people from another'. President Obama summarized US cultural approach to innovation when he said, 'Innovation does not just change our lives, it is how we make a living'. The City of San Diego has transformed itself over multiple decades into a strong vibrant innovation-based economy and a recent book by Walshok and Shragge, charting the journey has credited cultural values and social dynamics as being crucial in the transformation from a military-based economy to an innovation-based economy.

Peter Drucker often wrote that 'culture eats strategy for breakfast' and unless an organization or society's culture is open to innovation, it will not prosper. 'A starting point for the idea of openness is that a single organization cannot innovate in isolation' (Dahlander and Gann 2010). The 'openness' of an organization or society's culture to change and innovation is critically important to the creation and co-creation rate but particularly to the adoption of innovations. Countries such as Korea are known to have fast technology metabolism rates, which are conducive to rapid adoption of technology-based innovations. Organizations or countries which adopt Digital as a value or 'Digital First' as a mantra greatly increase the possibility of extracting optimized value from Digital technologies.

© Springer International Publishing Switzerland 2018
M. Curley, B. Salmelin, *Open Innovation 2.0*, Innovation, Technology, and Knowledge Management, DOI 10.1007/978-3-319-62878-3_13

## 13.1   Technology Metabolism Index

Dawn Mafus is an anthropologist at Intel who sought to understand why different countries adopt technologies at different rates with Korea and Estonia as standout example of fast adoption. Key factors that she identified are that both of these countries had agile governments, strong social networks, and major upheavals in living memory. These cases seem to be evidence of the old innovation axiom that 'necessity is the mother of invention'. Indeed Israel, which has more NASDAQ listed companies than the entire 27 countries of the USA, is a further example of this axiom at work.

Tuft's University and MasterCard have created a Digital Evolution Index (DEI) that tracks the progress countries have made in integrating digital technologies into the lives of their citizens and their economies. The DEI looks at factors across four key drivers, supply, demand, institutional environment, and innovation. Supply evaluates Internet access and infrastructure whilst demand explores consumer appetite for digital technologies. Government policies/laws and resources are the critical aspect of the institutional environment whilst the Innovation aspect examines investments into R&D and Digital Start-ups. The DEI classifies countries into four categories: standout, stall out, break out, and watch out. Stand out countries demonstrate high levels of digital development whilst also leading in innovation and new growth and in 2017 countries such as Singapore, the UK, UAE, and New Zealand were in this category. Stall out are countries which have a history of strong growth but have slowing momentum and this categories included many Western European countries. Break out countries such as Russia, Brazil, and Mexico are those that historically have had relatively low levels of digital advancement and are now demonstrating fast momentum. Watch out countries is a category where there are low levels of digital advancement and slow pace of growth. The index is published with a perspective that is very aligned with the OI2 ethos, providing a comprehensive report which allows different countries to learn from each other. The 2017 report also found that 'consumers' trust in digital technologies correlates with digital competitiveness', according to Bhaskar Chakravorti, Senior Associate Dean of International Business and Finance at The Fletcher School at Tufts University. Governments seeking to improve society and their economies can use the DEI as a way of learning best practice, influencing culture, and measuring progress.

## 13.2   Operational Excellence Versus Innovation Excellence

Many industrial organizations have their culture oriented towards supporting operational excellence and programs such as Lean Six Sigma are often in place to instil a quality and operational excellence mindset and discipline in the organization. However in modern organizations, cultures of operational and innovation excellence need to co-exist. The interesting thing is that the conditions needed to encourage and support innovation, for example, risk taking, experimentation, and tolerance of failure are exactly the opposite of those needed to support operational excellence, e.g. minimize deviations, no tolerance for failure. Creating the conditions for both these mindsets to exist in an organization or culture is difficult but important for sustainable success. One approach is to separate the innovation organization from the execution organization in an attempt to provide the right environment for disruptive innovation discovery and development and avoiding the tyranny of the urgent. When I was at Intel and we created the IT innovation centres, the then CIO Doug Busch deliberately separated the IT innovation centres from the rest of the organization so that the right culture and environment could exist for disruptive innovation.

Modern organizations are designed for operational excellence with the goal on variation minimization and risk reduction. However, the conditions needed for innovation are very different with requirement for risk taking and experimentation. An organization or society wishing to prosper needs to find a way for these two requirements to co-exist in harmony. Managing risk with open innovation and prototyping approach is one of the answers to the change drivers we see now. Serendipity must be given the chance.

## 13.3  High Expectation Entrepreneurship

According to Schumpeter, 'entrepreneurs' who develop new combinations of existing resources that accomplish innovation. High expectation entrepreneurship occurs when people with high ambitions collide with disruptive technologies. Clayton Christensen coined the term disruptive technologies, which describes a technology, which is initially inferior and cheaper than an incumbent technology but ultimately displaces the incumbent technology. The displacement of mainframe computers by the PC, or valves by transistors and integrated circuits are classic examples of disruptive technologies.

High expectation entrepreneurship deserves special focus because of its over-sized impact on economic growth. According to the Global Entrepreneurship Monitor, less than 7% of nascent entrepreneurs expect to employ fifty or more employees within 5 years; however, the economic impact is disproportionally positive as high-expectation entrepreneurs (HEEs) are responsible for up to 80% of total expected jobs by all entrepreneurs. Also as an example for a group of highly developed countries a 1% increase in the general rate of entrepreneurial activity raises economic growth by 0.11% while a 1% increase in high growth entrepreneurship yielded two-times multiplier effect with a 0.29% increase in GDP growth (Stam et al. 2007). Governments and policy makers should work to especially incentivize and lower barriers for HEEs. Ryanair is an example of a HEE, an airline which grew from one small regional airline to the largest carrier in Europe in just two decades, powered by a visionary leader with high ambition and demanding work ethics.

## 13.3.1  Openness to Innovation: Managing Six Vectors of Innovation

Machiavelli wrote 'There is nothing more difficult to take in hand, more perilous to conduct, or more uncertain in its success, than to take the lead in the introduction of a new order of things'. Thus for Innovation to prosper, there has to be an openness to innovation. In managing innovation, there are at least six vectors need to be managed successfully in parallel to ensure that the value promised through an innovation is delivered and realized.

These vectors are

- Vision
- Technology
- Business Ccse
- Business process change
- Organizational change
- Customer or societal change

Unless these vectors are managed in tandem, the chances of innovation success are significantly reduced. Ultimately, innovation only occurs when something is adopted and thus often the customer or societal change vector is ultimately the most important. People often resist change but when the benefits of change significantly outweigh the benefits of the status quo then adoption will happen. When a particular organization or society has an openness to change, then innovations are more likely to be adopted. One needs to be careful not to overload an organization or indeed a society with too much change as there is a limit to how much change individuals, organizations, and a society can absorb at one time.

> **Case Study: Alstom**
> Alstom is a French multinational operating worldwide in the railway business and previously in the energy business. Innovation is one of Alstom's strategic pillars and what makes the Group a world leader in its sectors. Innovation was a key factor for competitiveness, sustainability, and anticipating the customers' needs. To create new momentum in the Group's future technologies, Alstom created an Innovation Committee led by chief innovation officer Ronan Stephan. Five major cross-functional themes, as guidelines for R&D, were identified: materials and especially nanotechnologies, power electronics, digital, energy storage, and future urban systems. The innovation committee also fostered an 'open innovation' culture and strove to implement strategic partnerships with research teams, innovative start-ups, universities, and industry leaders in these fields. As part of this innovation culture, the Group organized an annual Innovation Awards whereby all Alstom employees were invited to submit innovative projects or inventions. These awards were introduced to foster a spirit of innovation in Alstom and stimulate the cross-fertilization of ideas. Leadership in culture has to come from the top and the clear prioritization of innovation as a priority in Alstom helped make it part of the DNA of the company.

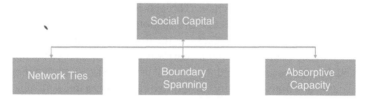

**Fig. 13.1** Connectedness, Enhancing the Innovation Capacity of an Organization (source: Mastrangelo)

## 13.4   Culture and Absorptive Capacity

Connectness and absorptive capacity of organizations in an ecosystem are crucial factors for success. Dolly Mastrangelo argues that 'that network connections across the ecosystem, combining deep expertise from different perspectives across industry, academia and government, will likely stimulate interactions to enhance innovation. These network interactions provide connectedness among people in enlightened organizations to leverage diverse networks of expertise and better address the new demands emerging to enhance value'.

Mastrangelo explains that there are four building blocks for connectedness (Fig. 13.1). They are as follow:

- Social capital
- Network ties
- Boundary spanning and
- Absorptive capacity

Mastrangelo defines these four building blocks as follows:

**SOCIAL CAPITAL**: the measure of connection amongst elements (people, business units, organizational entities, etc.). Social capital represents the willingness and motivation of people and organizations to interact with others.

**NETWORK TIES**: arrangements between people and organizations to interact. Ties can be strong or weak (each with benefits and drawbacks), formal or informal.

**BOUNDARY SPANNING**: is the act of crossing organizational boundaries to access external intellectual proper. These can be formal or informal with different levels of protection and accessibilities. Objective is to identify novel ideas.

**ABSORPTIVE CAPACITY**: a firm's ability to recognize the value of new information, assimilate it, and apply it to marketable ends.

Social capital refers to the networks of relationships amongst people who live and work in a particular industry, ecosystem, or society, enabling that industry, ecosystem or society to function effectively. The second building block is an organization's ability to build network ties to span boundaries with people in other organizations in the innovation chain. The crucial third building block of innovative capacity is spanning boundaries of other organizations to discover new external ideas across the innovation value chain. Through boundary spanning, individuals and organizations discover unexpected information to create new ideas for novel products or services. Mastrangelo argues that the first three building blocks are very interdependent as the ability to span boundaries is reliant on an organization's network ties and the culture as reflected in its social capital. The fourth building block is an organization absorptive capacity. Absorptive capacity is a measure of how well an organization recognizes and welcomes the value of external ideas, assimilates the ideas with current internal knowledge, and brings innovative products and services to market. Innovative organizations reflect on how their culture evolves and look to bring their connectedness to the appropriate level to optimize use of external networks and relationships. As indicated by Liu and Edvinsson in their work at New Club of Paris context, structural intellectual capital seems to have a crucial role when assessing and improving the competiveness of a country.

### 13.4.1   Social Innovation

Social innovation can refer to both social processes of innovation such as open source and peer production but more often than not, it refers to innovations, which have a social purpose. Organizations like Ashoka spearhead and support social innovation initiatives. Social innovation can refer to a broad set of outcomes such as new strategies, solutions, and organizations that meet and improve social needs such as public health to education to community development. Digital technologies can be a catalyst, platform, and resource in accelerating the development, diffusion, and adoption of social innovations. When social innovation and social media are tightly connected, there are great opportunities for fast, efficient, and effective impacts from social media.

## 13.5   Interdisciplinary Innovation

OI2 is also focused on interdisciplinary innovation and on what we call full spectrum innovation where the focus is not just on technical innovation but on equally important aspects such as business model innovation, process innovation, user experience, and brand innovation. Larry Keely's ten types of innovation model (Keeley et al. 2013) is a very useful taxonomy to help stimulate different kinds of innovation to maximize the possibility of successful adoption of an innovation.

OI2 subscribes to the philosophy that innovation is a discipline that can be practised by many rather than an art mastered by few and that successful innovations can become more predictable, probable, and profitable. The Innovation Value Institute has developed an innovation capability maturity framework and associated assessment tool, which helps guide organizations to improve their innovation capability and associated results. The goal is to deliver innovation results, which are more predictable, probable, and profitable.

OI2 requires a new mindset focused on teams, collaboration, and sharing. OI2 mindset combines courage and openness for innovation, but also courage to find disruptive solutions and let different disciplines ignite new ideas to be developed further.

Only with this focus will it be possible to tear down the walls that form separate silos of civil, academic, business, and government innovation. Silos will be replaced with creative commons, shared intellectual and societal capital, and the systematic harvesting of experimental results. Information and communications technology will play a special role because IT can supply the necessary connectivity/computing power and enable social networking amongst innovators and the communities they serve.

Lastly, OI2 subscribes to a broader view of growth aligned to a vision of sustainable intelligent living and aims to deliver a broader kind of value, one that Venkat Ramaswamy calls wealth, welfare, and well-being. In OI2, the goal of each innovation effort is not just to create wealth but also sustainable solutions which help improve welfare and well-being by solving key problems and seizing opportunities. With the continued development of information technology according to Moore's law, a key opportunity is to automate and dematerialize, substituting IT for other resources to deliver services which are better than prior services and are more resource efficient.

We believe the adoption of the new OI2 paradigm can be the catalyst that unleashes a virtual Cambrian explosion of innovation in Europe and beyond. Instead of gravitating to the lowest common denominator of its society, citizens will deliver to the highest common multiple by leveraging all the talents and resources of the broader society. Through co-innovation based on a shared vision and shared value creation, we can collectively move towards a vision of sustainable intelligent living. In the OI2 vision, each of us has multiple roles simultaneously in the innovation ecosystem. We have several professional roles, and several private roles in our knowledge sphere as professionals, in our experience sphere related to our family roles, citizen roles, in our free time hobbies, etc.

OI2 advocates for strong user involvement in the innovation process, and that co-creativity needs to be seamless in the process. Involving all stakeholders in a common shared vision means also that the users (citizens or user industries as example) are not anymore objects for innovation but active subjects in the process. In the new paradigm, users seamlessly take part in the ideation, development, and verification of the solutions in an iterative process, often based on real-time experiments and prototyping.

In the twenty-first century, the type of business models, depend on how technology is used, and instead of technology adapting to business models, business models have to adapt to technology. Technology includes open innovation and user-generated content. User-generated content has allowed companies such as Facebook, twitter, Rovio, an Apple to create their own benchmarks and become more competitive. This business model creates value by involving the customers in the creation process, while benefiting from their ideas.

New business models are key, but we should not forget the role of the 'society' in providing the safety net to protect the stakeholders. The role of the public sector will be changed with stronger empowerment of individuals and with unlocking the location dependence for many services. Policy will be derived from societal and technological experiments and prototypes as the changes to be expected are potentially quite radical. We cannot make policies by looking at the rear view mirror. The same applies to strategic business decisions. The need for early prototypes and experiments to create the 'new' either markets processes or products (including services) on strategic level has dramatically increased.

Often this user-enabled co-creation is dependent on available open engagement platforms. Apps platforms are good examples on how with relatively small cost experiments and scale-up of the successes are possible. Likewise, the Open Data can be a fruitful engine for many new applications for services.

In general, open innovation and the approach favours ecosystems with open engagement platforms for collaboration and creativity across all the stakeholders, and which lead to real-world experimentation and prototyping possibilities. Public sector has its clear role in driving for these platforms, in an open way. Competition will increasingly be between ecosystems and platforms rather than companies.

With the emergence of the Open Innovation 2.0 paradigm, there is an opportunity for a new entrepreneurial renaissance, which can drive a new wave of sustainable wealth creation. The aforementioned open platforms are also critical enablers to change the granularity of entrepreneurship from SMEs to even individuals who work outside traditional enterprises. Everyone can be a micro-entrepreneur and innovator.

## 13.6  Competitiveness of New Types of Organizations in Open Ecosystems

Technological revolution has a transformative nature. Society is moving towards a less hierarchical form, both in time and space. The 'power of the crowds', new business models conquering old models that were valid in the industrialized era, need to adjust to the new values. We have moved from an 'individual needs' basis to a 'societal needs' basis where the added value sometimes is more important than profitability. Even though, profits are made when value is added, the value is what makes the difference between competitive advantages and unstable businesses.

We have moved from an industrial, production incentive society to an ICT-based value generating society. And so, our values and basic needs have been adjusted to those higher up Maslow's hierarchy of needs. For example, we see that the levels of esteem and self-actualization start to grow through outlets such as social media. Therefore, the focus for successful innovation will be on the upper-level needs (Fig. 13.2).

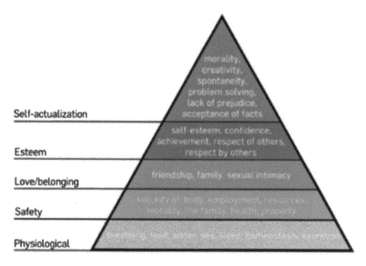

**Fig. 13.2**  Maslow's hierarchy of needs

**Fig. 13.3**  Adapted Maslow's hierarchy for the organization

A similar change arguably can be seen in enterprises. In the industrial era, the attention was cost oriented. In the current technological revolution, the most critical levels for success are cross-organizational issues, innovation culture, and the organization's agility to position the competences of the company in the society. Increasingly, the most successful organizations are those with a sense of purpose (Fig. 13.3).

A similar change can be seen in enterprises. In the industrial era, the attention was cost oriented. In the current technological revolution, the most critical levels for

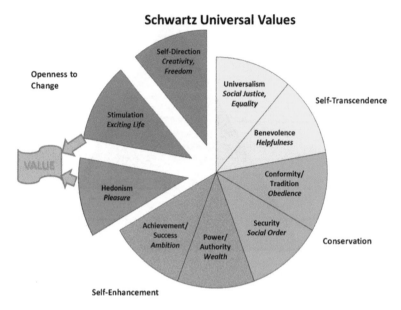

**Fig. 13.4**  Schwartz Universal Schwartz's Values Theory, Modified Pie Chart

success are cross-organizational issues, innovation culture, and the organization's agility to position the competences of the company in the society (Fig. 13.4).

Schwartz presented a captivating framework on the new values of entrepreneurs and enterprises. He used his 'Schwartz Value Inventory' (SVI) based on a wide survey of over 60,000 people to identify common values that acted as 'guiding principles for one's life'. He identified ten 'value types' that gather multiple values into a single category (9).

Schwartz stated that security and power were the main incentives for people at their jobs for many years. Many sociologists and psychologists believed, and some still do, that the motivation for a person to succeed in a professional environment is the ability to control others and to dominate resources paired up with safety and stability, and the comfort that these brings to their existence.

In relevance to innovation and 'today's principles', Schwartz presented a framework on 'new' values for innovation, entrepreneurship, and present-day enterprises and that is to create value. As we can see in the figure, hedonism and stimulation create tangible value and the environment for openness, creativity, and 'self-fulfilment'. Hedonists simply enjoy themselves. They seek pleasure above all things, which on its own may lead to debauchery that when paired up with stimulation gets a better mix. The need for stimulation is close to hedonism though the goal is slightly different. Pleasure comes from excitement and thrills. Additionally, Schwartz also presented and additional 'value' in the openness to change area, and it is self-direction, this value relates to those who enjoy being independent and outside the control of others. They prefer freedom and may have a particular creative or artistic bent; this creativity is what allows for innovation and entrepreneurship.

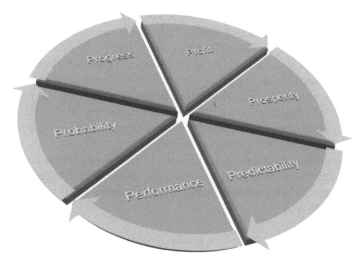

**Fig. 13.5** Output of OI2

Innovative personality traits naturally motivate individuals to be an entrepreneur. Self-direction, stimulation, and hedonism are not enough to become an entrepreneur, but they are enough to stir curiosity. In addition, the need for achievement, propensity to take risk, tolerance of ambiguity, self-confidence, and innovativeness are the traits argued by many other authors.

## 13.7  OI2 Outputs

The outputs of OI2 can be described by a number of different outputs profit, progress, prosperity, performance, probability, predictability, and payback–return on innovation. All of these different outputs are intertwined, some are very short term and others are very long term. We all know that innovation is risky but our hypothesis is that by using the OI2 patterns, the probability of innovation success using platforms and ecosystems can significantly improve and that multiplicative effects can be achieved. The win-win outcomes that be achieved by using OI2 take us closer to what Stahel calls the performance cconomy. Ultimately, however, money talks so we shall briefly discuss payback and return on innovation as this is the language that financiers and accountants mostly understand (Fig. 13.5).

## 13.8  Payback–Return on Innovation

Innovation is risky and so anything can be done to improve the probability of innovation success is very worthwhile. Sustainable innovation is even more difficult and elusive. A closer examination of US stock market returns, which has been performed by Bessembinder (2017), is very revealing. Over a ninety year period only thirty companies (out of over twenty five thousand companies) accounted for one third of the cumulative wealth of the entire US stock market. Half of the wealth created over that 90-year period was contributed by just 0.33% of all companies. In addition, less than 1.1% of the stocks that existed in that period contributed three quarters of the stock market's cumulative gains. The top one thousand performing companies, less than 4% of listed companies, since 1926 have accounted for all the stock market gains. The corollary of this is that for 96% of these companies investors would have been better off investing in Government treasury bills.

By using the value dials approach explained in the shared purpose pattern, prospective innovations can be defined, designed, and delivered against targeted financial improvements. When network effects are involved supra-linear and exponential returns can be achieved.

## 13.9   OI2 Outcomes

In parallel with improved output, OI2 delivers improved outcomes as shown in the following figure (Fig. 13.6).

By working together using OI2, we get a range of outcomes including alignment of innovation intent, amplification of resources, acceleration of results due to the power of the ecosystem, risk attenuation because of shared risk, agility as we can leverage the power of the ecosystem and co-innovation on a common platform, and finally action as we have latent energy converted into potential energy through action. The collective impact of these outcomes is that the efficiency and effectiveness of innovation efforts can be significantly improved as an ecosystem will always

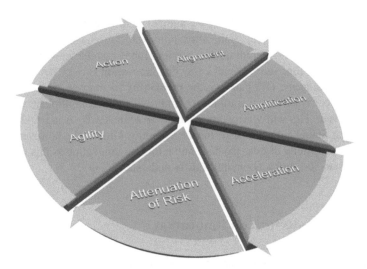

**Fig. 13.6**  OI2 Outcomes

have economies of scope and scale advantages over individual companies. This is really the power of OI2, with organizations and individuals working together to accelerate results which they could not hope to deliver on their own. Whether for an ecosystem or for a societal level system such as an energy grid or a transportation system, the OI2 patterns can help improve the predictability and probability of results as well as their shared profitability.

## 13.10   Fast and Slow Innovation Cycles

Innovation requires co-creation and participation by all stakeholders in their complementary roles. In the figure below, we see two innovation cycles; the rapid one (1) which builds on fast user/co-creator feedback in real-world settings (Fig. 13.7). Here, the users together with the industry are co-creating new services and products,

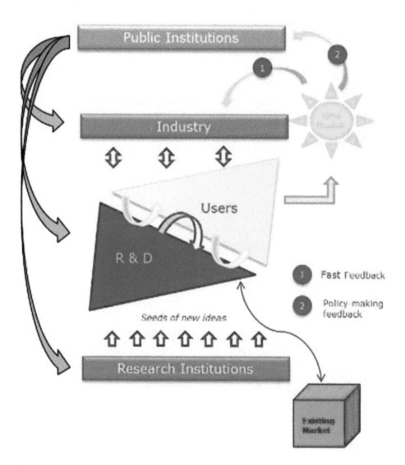

**Fig. 13.7**   Fast and Slow innovation cycles

and even more, markets together. The R&D seed is filtered through user experience and needs, and opportunities as well to experimentations and prototypes in the new markets to see what flies and what is to be left out. This iterative process is dynamic and has a lot of serendipity in it by nature.

The slower innovation cycle (2) is related to innovation infrastructures and framework creation for frictionless and rich innovation ecosystems. The public institutions have here an important role by focusing public procurement for innovation thus scaling up innovations coming from the user-industry-research collaboration. Public sector is an important buyer who is able to change the innovation landscape significantly. Public research and infrastructure funding, together with creating the right framework conditions (like IPR) is influencing the richness and fluidity of the innovation ecosystem directly.

Public funding enables research institutions to invest in fundamental explorative research, growing seed for the R&D to harvest into innovation injection. It is very important to view these two cycles separately as they involve same players but in different interconnected roles. In addition, the innovation policy actions should take these both into account to have a balanced short-term and long-term perspective on the innovation ecosystem development.

# Chapter 14
# Looking Forward

According to Steven Carter, author of 'Where good ideas come from?' the great driver of scientific and technological innovation has been the historic increase in connectivity. Indeed, we are witnessing what Kurzweil called the law of accelerating returns with each new innovation building on prior innovations and also often becoming infrastructure for future innovations. The OI2 innovation paradigm is based on extensive networking and co-creative collaboration between all actors in society, spanning organizational boundaries well beyond normal licensing and collaboration schemes. With the ever-increasing speed and rate of connectivity, we are creating a giant global intellectual supercollider and the possibility exists that people and machines alike may participate in a giant neural network, which spans the globe. Maybe this is what Peter Russell had in mind when he wrote the Global Brain book in the 1970s.

Looking forward, we need to collectively adopt the mindset of shared vision and shared value, and build innovation strategies and ecosystems to tackle the major societal problems. For example, we could build the equivalent of Moore's law and an ecosystem to deliver healthcare transformation—systematically and continuously finding technology interventions which will improve quality of life and quality of care which when cumulatively added help create longer healthier lives—a key role of the citizen would be taking more individual responsibility for their health. In parallel, there will be difficult challenges to solve in areas such as security, standards, trust, and privacy as more and more systems are open and interconnected. However, this should not stop us making progress. Tactically, the EU and government agencies in approving research projects for public funding should take into account criteria like on 360 participation of actors and the quality of the proposed design pattern(s) to help successfully diffuse the outcomes of the project.

Ultimately, it is not just about open innovation, but openness to innovation. Again, Peter Drucker wrote that culture eats strategy for breakfast everytime, highlighting the importance of culture to the success of any strategy. Fostering a culture, which is open to innovation and risk taking, is important. Increasingly, it seems

© Springer International Publishing Switzerland 2018
M. Curley, B. Salmelin, *Open Innovation 2.0*, Innovation, Technology,
and Knowledge Management, DOI 10.1007/978-3-319-62878-3_14

there is increasing appetite by citizens to be involved in larger innovation efforts, as exemplified by more citizen science initiatives. In a Dublin City Council survey of visitors to a future cities exhibition in Dublin in 2013, over 90% of respondents felt that Dublin should be used as a venue for testing experimental solutions and they would be willing to participate in the experiments themselves. The UK government's whitepaper on community engagement 'Communities in control: real people, real power' also indicates that there is a desire from the top for more community and personal engagement.

Denis and Donatella Meadows et al. wrote in 1972 that 'Man Possesses, for a small moment in time, the most powerful combination of knowledge, tools and resources the world has ever known. He has all that is physically necessary to create a totally new form of human society—one that would be built to last for generations. The missing ingredients are a realistic long term goal that can guide mankind to the equilibrium society and the human will to achieve that goal'.

It is strange that this statement seems even truer today and yet the progress made has been disappointing. We have the technology and now we now have an emerging innovation paradigm and methodology. We have to take the opportunity of a lifetime, in the lifetime of this opportunity. The technology is almost ready—are we? Let's create our future together with the OI2 approach!

# References

Aho E (2006) Creating an innovative Europe. EU Publications, Luxembourg

Alexander C, Ishikawa S, Silverstein M (1977) A pattern language: towns, buildings, construction. Oxford University Press, Oxford

Amidon D, Formica F, Mercier-Laurent E (2004) Knowledge economics: emerging principles, practices and policies. University of Tartu, Tartu

Andersson T, Formica P, Curley MG (2009) Knowledge-driven entrepreneurship: the key to social and economic transformation. Springer, Berlin

Andersson T, Formica P, Curley M (2010) Knowledge Driven Entrepreneurship, the key to social and economic transformation. Springer, New York

Appleton B (2000) Patterns introduction. Available at http://www.sci.brooklyn.cuny.edu/~sklar/teaching/s08/cis20.2/papers/appleton-patterns-intro.pdf

Assimakopoulous DG, Oshri I, Pandza K (2015) Managing emerging technologies for socio-economic impact. Edward Elgar, Cheltenham

Baldwin E, Curley M (2007) Managing IT innovation for business value. Intel Press

Bass F (1969) A new product growth for model consumer durables. Manag Sci 15(5):215–227

Bessembinder H (2017) Do stocks outperform treasury bills? (18 Feb 2017). Available at SSRN: https://ssrn.com/abstract=2900447

Brunswicker S, Hutschek U (2010) Crossing horizons: leveraging cross-industry innovation search in the front-end of the innovation process. International Journal of Innovation Management 14(4):683–702

Brunswicker S, Vanhaverbeke W (2015) Open Innovation in Small and Medium-Sized Enterprises (SMEs): external knowledge sourcing strategies and internal organizational facilitators. J Small Bus Manag 53:1241–1263

Carayannis EG, Campbell DFJ (2009) "Mode 3" and "Quadruple Helix": toward a 21st century fractal innovation ecosystem. Int J Technol Manag 46(3/4):201–234. http://www.inderscience.com/browse/index.php?journalID=27&year=2009&vol=46&issue=3/4

Carayannis EG, Campbell DFJ (2010) Triple helix, quadruple helix and quintuple helix and how do knowledge, innovation and the environment relate to each other? A proposed framework for a trans-disciplinary analysis of sustainable development and social ecology. Int J Soc Ecol Sustain Dev 1(1):41–69. http://www.igi-global.com/bookstore/article.aspx?titleid=41959

Carayannis EG, Campbell DFJ (2011) Open Innovation Diplomacy and a 21st Century Fractal Research, Education and Innovation (FREIE) ecosystem: building on the quadruple and quintuple helix innovation concepts and the "Mode 3" knowledge production system. J Knowl Econ 2(3):327–372

Chesbrough H (2003) Open innovation: the new imperative for creating and profiting from technology. Harvard Business School Press, Boston

COP21 United Nations Conference on Climate Change. http://www.cop21.gouv.fr/en

Curley M (2004) Managing information technology for business value. Intel Press, Hillsboro

Curley M (2013) Plan C for a digital low-carbon future – less is moore, industry foreword, G8 climate change – New Economy

Curley M (2015) The evolution of innovation. The Journal of Open Innovation, Luxembourg

Curley M (2016) Twelve principles for Open Innovation 2.0. Nature. Accessed at http://www. nature.com/news/twelve-principles-for-open-innovation-2-0-1.19911

Curley M, Formica P (2013) The experimental nature of new venture creation, capitalizing on Open Innovation 2.0. Springer, London

Curley M, Salmelin B (2013) Open Innovation 2.0—a new paradigm (EU Open Innovation and Strategy Policy Group). OI2 Conference Paper

Curley M, Salmelin B (2014) Open Innovation 2.0 – a new paradigm, in Open Innovation 2.0. EU Publications

Curley M, Kenneally J, Ashurst C (2012) Design Science and Design Patterns: a rationale for the application of design-patterns within design science research to accelerate knowledge discovery and innovation adoption. EDSS proceedings, Springer

Curley M, Kenneally J, Carcary M (2016) IT Capability Maturity Framework (IT-CMF): the body of knowledge guide. Van Haren, Netherlands

Dahlander L, Gann DM (2010) How open is innovation? Res Policy. doi:10.1016/j. respol.2010.01.013

Dahlander L, O'Mahony S, Gann DM (2014) One foot in, one foot out: how does individuals' external search breadth affect innovation outcomes? Strateg Manage J

De Bresson C, Amesse F (1991) Networks of innovators: a review and introduction to the issue. Res 20:363–379

Dougherty D, Dunne D (2011) Organizing ecologies of complex innovation. Organ Sci 22(5):1214–1223

EU Energy Union (2015) http://europa.eu/rapid/press-release_MEMO-15-4485_en.htm

EU Open Innovation and Strategy Policy Group (OISPG) (2010–2016) Annual Open Innovation 2.0 yearbook

Fowler M et al (2002) Patterns of enterprise application architecture. Addison Wesley, Boston

Gamma E, Helm R, Johnson R, Vlissides J (1994) Design patterns: elements of reusable object-oriented software. Addison Wesley, Boston

Gawer A, Cusumano MA (2013) Industry platforms and ecosystem innovation. Wiley J Prod Innov Manag 31(3): 417–433

Global Footprint Network (2015) http://www.footprintnetwork.org/en/index.php/GFN/page/world_footprint/. Accessed 18 Sept 2015

Greenwood Mastrangelo D (2013) Collaborate to innovate: innovative capacity for effective open innovation. Doctoral Thesis

Hansen M, Birkinshaw J (2007) The innovation value chain. Harvard Business Review, June 2007

Haque U (2011) Betterness: economics for Humans. Harvard Business Press, Cambridge

Holt K (2002) Market oriented product innovation. Springer, Dordrecht

Hwang V (2012) What's the big deal about Innovation Ecosystems. Forbes

Johansson F (2006) DeMedici effect: what elephants and epidemics can teach us about innovation. Harvard Business Review Press

Keeley L, Walters H, Pikkel R, Quinn B (2013) Ten types of innovation: the discipline of building breakthroughs. Wiley, Hoboken

Kuhn TS (1962) The structure of scientific revolutions. University of Chicago Press, Chicago

Laitner, Erhardt-Martinez K (2008) Information and communication technologies, the power of productivity. ACEEE, Washington, DC

Leiponen A, Helfat CE (2011) Location, decentralization, and knowledge sources for innovation. Org Sci 22(3):641–658

Liu J (2016) Industrial mashups, EY. http://www.ey.com/Publication/vwLUAssets/ey-industrial-mash-ups/%24FILE/ey-industrial-mash-ups.pdf

Madelin R (2016) EU innovation report. European Commission, Luxembourg

Marjanovic S, Fry C, Chataway J (2012) Crowdsourcing based business models: in search of evidence for innovation 2.0. Sci Public Policy 39(3):318–332

McInerney F, White S (2000) Future wealth: investing in the second great wave of technology stocks. Saint Martin's Press, New York

Metcalfe R, Boggs D (1976) Ethernet: distributed packet switching for local computer networks. Commun ACM 19(7):395–405

Meyer MH, Lehnerd AP (1997) The power of product platforms: building value and cost leadership. Free Press, New York

Nambisan S, Nambisan P (2013) Engaging citizens in co-creation in public services: lessons learned and best practices. IBM Center for the Business of Government, Washington, DC

Obama B (2009) A strategy for American innovation: driving towards sustainable growth and quality jobs. report from the White House, Washington, DC

OECD (2010) Ministerial report on the OECD Innovation Strategy. Innovation to strengthen growth and address global and social challenges

Osterwalder A, Pigneur Y (2010) Business model generation: a handbook for visionaries, game changers, and challengers. Wiley, Chichester

Pham P (2015) Why you may need to change the way you look at big data. Forbes. https://www.forbes.com/sites/peterpham/2015/09/30/why-you-may-need-to-change-the-way-you-look-at-big-data/#3eda911140a4

Pisano G, Verganti R (2008) What kind of collaboration is right for you? Harvard Business Review

Porter ME, Kramer MR (2011) Creating shared value. Harv Bus Rev 89(1/2):62–77

Ramaswamy V, Goiullart F (2010) The power of cocreation. Free Press, New York

Ramaswamy V, Ozcan K (2014) The Co-creation paradigm. Stanford University Press, Redwood City

Rogers EM (1962) Diffusion of innovations. Free Press, Glencoe

Russell P (1983) The Global Brain: speculations on the evolutionary leap to planetary consciousness. Los Angeles: JP Tarcher

Ryan J (2010) A history of the internet and the digital future. Reaktion Books, London

Sarsgyan MG (2013) User driven innovation. EU Open Innovation 2.0 yearbook 2013

Schrage M (2004) Michael Schrage on innovation. Interview in ubiquity, a peer-reviewed Web-based magazine. ACM Publication, December ubiquity.acm.org

Schumpeter JA (1942) Capitalism, socialism, and democracy. Routledge, London

Stahel W (2010) The performance economy, 2nd edn. Palgrave Macmillan, London

Stam E, Suddle K, Hessels J, Van Stel A (2007) High growth entrepreneurs, public policies and economic growth. Ekonomiaz, Basque Journal of Economics – Special issue on entrepreneurship

Tapscott D, Williams AD (2006) Wikinomics: how mass collaboration changes everything. Penguin, New York

Technology CEO Council (2008) A smarter shade of green, Washington. Accessed at http://www.techceocouncil.org/news/2008/02/06/press_releases/new_study_links_technology_to_dramatic_energy_efficiency/

Tiwana A (2014) Platform ecosystems, aligning architecture, governance and strategy. Morgan Kaufman, San Francisco

Tuckman BW (1965) Developmental sequence in small groups. Psychol Bull 63(6):384–399

UN (2015) Transforming our world: the 2030 Agenda for Sustainable Development, A/RES/70/1 United Nations

UNEP (2011) Decoupling natural resource use and environmental impacts from economic growth. International Resource Panel

Vaishnavi V, Kuechler W (2004/2005) Design research in information systems, 20 Jan 2004; last updated 15 Nov 2015. http://desrist.org/design-research-in-information-systems

Vallat J (2010) The IP implications of Open Innovation. EU OISPG

Von Hippel E (1988) The sources of innovation. Oxford University Press, Oxford

Walshok M, Shragge A (2014) Invention and reinvention, the evolution of San Diego's innovation economy. Stanford University Press, Stanford

World Commission on Environment and Development (1987) Our common future. Oxford University Press, Oxford. p 27. ISBN 019282080X. this is also known as the (Brundtland report)